THE
FOURTH DIMENSION
SIMPLY EXPLAINED

THE
FOURTH DIMENSION
SIMPLY EXPLAINED

A Collection of Essays Selected from
Those Submitted in the
Scientific American's Prize Competition

Edited by
HENRY P. MANNING

late Professor of Mathematics in
Brown University

Introduction to the Dover Edition by
THOMAS F. BANCHOFF
Brown University

DOVER PUBLICATIONS, INC.
Mineola, New York

Bibliographical Note

This Dover edition, first published in 1960 and republished in 2005, is an unabridged republication of the work originally published by Munn & Company, New York, in 1910. Thomas F. Banchoff has written a new Introduction specially for the 2005 edition.

Library of Congress Cataloging-in-Publication Data

The fourth dimension simply explained : a collection of essays selected from those submitted to the Scientific American's prize competition / edited by Henry P. Manning ; introduction to the Dover edition by Thomas F. Banchoff.
    p. cm.
  Originally published: New York : Munn & Co., 1910.
  Includes bibliographical references.
  ISBN 0-486-43889-9 (pbk.)
  1. Fourth dimension. I. Manning, Henry Parker, 1859–

QA699.M3 2005
530.11—dc22

2004059330

Manufactured in the United States of America
Dover Publications, Inc., 31 East 2nd Street, Mineola, N.Y. 11501

INTRODUCTION TO THE
DOVER EDITION

BY THOMAS F. BANCHOFF, BROWN UNIVERSITY

IN 1909, *Scientific American* ran a contest challenging readers to answer the question "What Is the Fourth Dimension?" in 2500 words or less. The results were surprising: 245 entries from all over the world, all seeking the $500 prize–quite a substantial inducement in those days. It was easy to find a judge for the contest: Henry Parker Manning (1859–1956) had returned to his alma mater, Brown University, after receiving a Ph.D. from the preeminent research university, Johns Hopkins, in 1891, and for the next twenty-nine years he taught all of the advanced courses, including Theory of Substitutions, Complex Analysis, and, most unusual for that time, Higher-Dimensional Geometry and Quaternions. He had already published an exposition of non-Euclidean geometry with a historical introduction, and he was well familiar with the literature in Europe as well as the U.S.

Things moved quickly in 1909. The deadline for entries was in April, and the prize-winning essay and the three honorably mentioned entries were published in the *Scientific American* issues in July of that year. Manning was asked to edit a selection of essays for publication in a separate volume, and it appeared the next year, with a lengthy introduction and frequent footnotes.

Most modern readers are quite familiar with time being a fourth dimension, and in fact they are surprised that there are any other ways to think about four dimensions. It is therefore doubly surprising to see that not one of the twenty-six essays in this collection considers time as a fourth dimension. Although Albert Einstein had announced his theory of Special Relativity in 1905, this idea had not yet filtered down to the public consciousness four years later. When Manning wrote his book *Geometry of Four*

Dimensions in 1914, he mentioned that "Another very important application of geometry of four dimensions is that mentioned by d'Alembert, making time the fourth dimension: within a few years this idea has been developed very fully, and has been found to furnish the simplest statement of the new physical principle of relativity," with a footnote "This important application of four-dimensional geometry was developed by Minkowski."

Five years earlier, however, everyone was concerned with a fourth dimension of space rather than time, almost always for one of two reasons: the curious challenges of extending ordinary geometry from the plane to three-dimensional space and beyond, or the fascination or repulsion of the ideas of the spiritists—believers in parapsychological phenomena that they were certain had their origin in a fourth-dimensional spirit world. Almost all of the essayists explored analogies for constructing a four-dimensional geometry that extended our understanding of ordinary three-dimensional space, and almost all of them debunked the spiritist claims. Manning explains that there is a great deal of repetition in the essays, although there are differences in emphasis that justify their selection.

The Dimensional Analogy

For several years after his graduation from Brown University in 1883, Manning taught in secondary schools, so he was aware of the importance of the dimensional analogy in the curriculum. He starts his introduction "The geometry studied in the schools is divided into two parts, Plane Geometry, or Geometry of Two Dimensions, and Solid Geometry, or Geometry of Three Dimensions, and the study of these geometries suggests an extension to geometry of four or more dimensions." Unfortunately a good deal of solid geometry has disappeared from the secondary-school curriculum, so the power of the dimensional analogy is no longer so strong. This may be one of the most important reasons to republish these essays, reflecting as they do a common appreciation of the geometrical background of the average reader of *Scientific American* nearly a century ago.

Spiritism and the Fourth Dimension

Also unfamiliar to most modern readers is the attention paid to the claims of the spiritist in America and in Europe in the latter part of the nineteenth century. According to Carl Cranz, writing in 1890, "Modern spiritism must have 8–11 million adherents, primarily in the formative stages, who put down their results and points of view in 25 periodicals and hundreds of other independent works." Perhaps the most famous figure in the spiritist story was Johann Carl Friedrich Zoellner (1834–1882), a famous and respected professor of physical astronomy at the University of Leipzig, who became convinced that it was possible to prove the existence of a world of four dimensions. He was taken in by the sleight of hand of an American magician and confidence man, Henry Slade, who claimed to be able to perform feats only possible by having access to the fourth dimension, for example untying a knot in a closed loop of string or removing the contents of a chest without breaking the seal. Although professional magicians exposed his claims as crude tricks, Zoellner staked his reputation on this evidence for a physical fourth dimension and became discredited by the scientific community of his day. He died at age 48 in 1882 but his story was still quite current in the minds of the essayists, primarily because of popular articles by several writers mentioned in more than one essay.

Articles Cited

Hermann Schubert (1848–1911), writing in *The Monist* in 1903, attacked Zoellner's gullibility in his article "The Fourth Dimension," and the article was reprinted in the popular volume *Mathematical Recreations and Essays.*

Charles Howard Hinton (1853–1907) wrote his article "The Fourth Dimension" in *Harper's Magazine* in 1904, and several of his books on the topic were widely known. Although several essayists refer to Hinton's writings, not one of them cites the most enduring book about the fourth dimension, namely *Flatland* by Edwin

Abbott Abbott (1838–1926). Manning specifically compares the two authors in a footnote in his introduction: "Such a book [about life in a two-dimensional world] has recently ben written by C. H. Hinton: 'An Episode of Flatland.' But much better is a little book by E. A. Abbot [sic] called 'Flatland.' There the interest rests entirely on the relations of space which this book is intended to explain, and we never for a moment lose sight of these relations."

Simon Newcomb (1835–1909), who died the year of the *Scientific American* contest, was the most famous astronomer of his day, a professor at Johns Hopkins University, where Manning received his Ph.D. in 1891. Newcomb wrote an essay, "The Fairyland of Geometry," in *Harper's Magazine* in 1902, a reworking of his retiring presidential address before the American Mathematical Society in 1897. One of the surprising things to a modern reader is his description of the "lumeniferous ether" as the context against which physics takes place. The Michelson-Morley experiment took place in 1887 but its full significance had not had an impact on the ether theory. Karl Pearson, one of the scientists cited in an honorable-mention essay, developed a theory of atoms as "squirts in the ether." Newcomb was one of Manning's professors at Johns Hopkins University.

The Essays and the Essayists

The prize-winning essay was by Lt. Col. Graham Denby Fitch, a graduate of West Point who served in the Corps of Engineers in the Spanish-American War and spent his retirement on river and harbor improvement projects. His only publication, other than his essay for the contest and another written specifically for this volume (Essay II—"Non-Euclidean Geometry and the Fourth Dimension") was "Lock and Dam Construction on the Upper White River, Arkansas." He was 49 at the time of the competition.

A number of other entrants had careers in mathematics and science.

George Gailey Chambers (1873–1935; Essay XVIII—"The Meaning of the Term 'Fourth Dimension' ") had just received his Ph.D. in mathematics from the University of Pennsylvania in 1908 with the thesis "The Groups of Isomorphisms of the Abstract Groups of Order p^2q." He served as an alumni trustee of Dickinson College. In his essay he refers to another Penn graduate student, Paul R. Heyl, who constructed models of projections of figures in four-space and whose Ph.D. thesis was entitled "Theory of Light on the Hypothesis of a Fourth Dimension."

Arthur Robert Crathorne (1873–1946; Essay XIII—"The Fourth Dimension, the Playground of Mathematics") received his Ph.D. in 1907 from the Georg-August-Universität, Göttingen, with a thesis called "Das räumliche isoperimetrische Problem" (The Isoperimetric Problem in Space) with the great David Hilbert as his thesis advisor. He taught mathematics and statistics at the University of Illinois from 1907 until 1946.

By coincidence, Crathorne founded the Institute of Mathematical Statistics in Michigan in 1935 along with another contestant, Burton Howard Camp (Essay IX—"The Fourth Dimension Algebraically Considered"), who received his Ph.D. from Yale in 1911 with the thesis "The Convergence of Singular Integrals." Camp went on to teach at Wesleyan University and wrote *The Mathematical Part of Elementary Statistics*, which was published in 1931.

Mrs. Elizabeth Brown Davis (1863–1917; Essay X—"Difficulties in Imagining the Fourth Dimension") "graduated with a B.S. from Columbian University (now George Washington) in Washington, D.C. She did her post-graduate work at Johns Hopkins University in mathematics by special permission of the faculty through Prof. Simon Newcomb. She became a computer [!] in the United States Naval Observatory and assisted in the preparation of Newcomb's Tables of the Sun and Planets."[1] Davis's last published paper was "Relating to an Extension of Wilson's Theorem" in the *American Mathematical Monthly* in 1917.

---

[1]From the U.S. Naval Observatory's website,
maia.usno.navy.mil/women_history/davis.

Arthur Erich Haas (1884–1941) (Essay VII–"An Interpretation of the Fourth Dimension") was a physicist who taught in Vienna, Leipzig, London, and, from 1936, at the University of Notre Dame. "After first doing work in the history of science, he turned in 1909 to theoretical physics and made a number of proposals in the field of atomic physics, anticipating Bohr's theory of the atom."[2]

John Clyde Hostetter (Essay XX–"Possible Movements and Forms in a System of Four Dimensions") wrote unter the pseudonym "Der Chemiker." He was an Alumni Trustee of Bucknell University (1939–1944). He specifically mentions the mirror-symmetric forms of tartrate crystals (page 230), the polarized light properties of which were discovered originally by Louis Pasteur. He is the only essayist who acknowledges any outside references, specifically Hinton and Schubert, "whose papers have been freely used in the preparation of this essay."

Leonard Gunnell (1869–c.1946; Essay VIII–"Length, Breadth, Thickness, and Then What?") was librarian of the Smithsonian Institution.

George Moreby Acklom (Essay XI–"Some Fourth-Dimensional Curiosities") was born in 1870 to English parents in India and ran a prestigious boy's school in Halifax, Nova Scotia, before moving to New York City in 1907. He was a literary advisor for the E. P. Dutton Publishing Co., and a poem of his was included in a collection of Canadian sonnets published in 1910. Acklom's wife was a relative of the woman who became Princess Diana, and their son–who took the stage name David Manners–was a Hollywood star whom both Lucille Ball and Marlon Brando credited with giving them their breaks in the film industry.

A. Clement Silverman (Essay XXI–"The Fairyland of the Fourth Dimension") was a pediatrician from Syracuse who starts his essay by suggesting that "perhaps we, too, are but babies in a space of a fourth dimension, and that we, too, might

---

[2]From the Columbia Electronic Encyclopedia, reference.allrefer.com/encyclopedia/h/haas-art.

be no less astonished if beings from that world chose to play peek-a-boo with us."

The essayist who did the most subsequent work on the fourth dimension was Claude Fayette Bragdon (1866–1946; Essay VI–"Space and Hyperspace"), an architect, stage designer, and writer whose works include *Projective Ornament* (1913) and *Four-Dimensional Vistas*(1916). In his biography, *More Lives Than One*, he neglects to mention his contest entry. He was the translator and popularizer of the influential Russian work with a four-dimensional theme, *Tertium Organum* by P. D. Ouspensky. Bragdon's 1928 article "The Fourth Dimension," written as an accompaniment to Ouspensky's book, concludes "Dimensionality is the mind's method of mounting to the idea of the infinity of space. The fourth dimension is the fourth stage in the apprehension of that infinity."

Henry Parker Manning

In his Introduction, Henry Parker Manning makes a case for studying the fourth dimension which explains in a large measure his choice for the prize-winning essay. Manning and Lt. Col. Fitch think alike, and they present their ideas in much the same style. Both of them revere geometry both as an abstract system based on axioms and as world of shapes that can be described visually. Both writers are guided by the process of analogy, not for proving that some system or other corresponds to physical reality, but for suggesting mathematical ideas in higher dimensions that are as real as the two- and three-dimensional phenomena from ordinary Euclidean geometry. Both of them prefer a style where results are listed without any attempt either at formal proof or potential application to the physical world.

At several places, Manning introduces ideas that resonate very well with ideas that are current in mathematics and in mathematical physics. When he describes a "material surface" as "a substance having two dimensions of considerable extent and two dimensions small" he is very close in spirit to modern string theory. In the fourth section of his Introduction, which he suggests that his readers only look at after they have read some essays, he spends a good deal of time discussing the analogues of wheels, as well as figures in four-dimensional space that correspond to cones and cylinders. The figure that intrigues him the most is something that is hardest to visualize, namely a "double cylinder" formed by moving a disc around the boundary curve of another disc, and then doing the reverse process with the roles of the disc interchanged. In modern terms, Manning is describing the decomposition of the three-dimensional sphere (the boundary of a four-dimensional ball) as the union of two solid torus-shaped surfaces joined along a common-boundary torus surface. It took modern computer graphics to provide a way of visualizing such a decomposition, using central projection to show a surface in our space that separates all of three-dimensional space into two congruent parts. Moving the surface one way gives one solid torus, and moving another way gives the other. Manning would have appreciated the power of interactive computer graphics to investigate some of his favorite figures in a space of four dimensions.

Manning continued his interest in higher-dimensional geometry, producing materials on quaternians (as they were called before the modern spelling quaternions was established), and, as already mentioned, a book, *Geometry of Four Dimensions,* in 1914. He became interested in relativity and delivered popular lectures on that subject (as recounted by the president of the Beta Theta Pi fraternity at Brown University). He also produced small books on *Non-Euclidean Geometry* and *Irrational Numbers and Their Representation by Sequences and Series,* both related to his advanced courses on these subjects. One of his students was Albert A. Bennett, who later wrote the review of William Anthony Granville's 1922 book *The Fourth Dimension and the Bible* for the *American Mathematical Monthly* (and who was still at Brown when I first arrived there in 1967). Manning contributed to David Eugene Smith's 1929 *Source Book in*

Mathematics. In his letters to Smith, available in the Columbia University Library, he expresses his confidence in the subject of higher dimensions, and he offered articles on several of the most important mathematicians working in that area.

Manning stopped teaching in the early 1920s and retired from the mathematics department in 1926. His students remembered him fondly, and a number of them sent in reminiscences for a ceremony in 1983–the centennial of his graduation from Brown University–dedicating the Henry Parker Manning seminar room in Brown's mathematics department. The characteristic they most often recalled was that he was almost completely deaf, so that when he turned to write on the blackboard, students were sure that he could not know that they were talking aloud among themselves. But Professor Emeritus H. Bruce Lindsey, remembering his own student days, recalled that Prof. Manning sent a note to him and another student saying that he knew what happened when his back was turned: "I may be deaf but I'm not stupid." The two students atoned for their misbehavior by keeping the others in line for the rest of the year.

Several friends recall his times in Weld, Maine, where the Mannings would go for their vacations. He always had work spread out on a table in the cabin, and because he was deaf, he could continue to work even when conversations were going on. Frances Wright, professor of astronomy at Harvard University, recalls his help with mathematics during those summers.

In his retirement, Manning worked closely with his friend Chancellor Arnold B. Chace (who had graduated from Brown in chemistry in 1866 and received an honorary doctorate from Brown in 1892), who was very involved with mathematics, attending most of the advanced classes taught by Manning. Chace visited Harvard University to work with Prof. Benjamin Peirce (leading to an article involving quaternions in the *American Journal of Mathematics*), and in 1910 became very interested in the British Museum's Rhind Papyrus, whose translation and publication he undertook through the Mathematical Association of America. Manning was responsible for the mathematical interpretations, which he did with some success, as reported by his colleague Raymond Clare Archibald in his testimony at Manning's retirement. Letters at Brown University from scholars in Egyptology praise Manning's contribution to the field, including the solution of some perplexing items in the papyrus. One letter is from Prof. Otto Neugebauer at the University of Göttingen, later a distinguished professor of the history of mathematics at Brown. He apologized to Manning that he was writing in German because he did not feel confident discussing these subjects in English! (When I arrived at Brown, Neugebauer was our most distinguished professor, and spoke English very well and confidently–along with being confident in a spate of other modern languages and half a dozen not spoken for at least 5000 years.)

Despite his record as a teacher and despite his record for publication of historical articles and graduate-level expositions on non-Euclidean geometry and irrational numbers, Manning never became a full professor during his active time as a faculty member at Brown. Only in 1946 was the title Full Professor Emeritus conferred on him by the Brown administration. At the time of his death in Providence in 1956, he was the oldest living member of the American Mathematical Society.

The
FOURTH DIMENSION
Simply Explained

PREFACE

IN January, 1909, a friend of the Scientific American, who desired to remain unknown, paid into the hands of the publishers the sum of Five Hundred Dollars, which was to be awarded as a prize for the best popular explanation of the Fourth Dimension, the object being to set forth in an essay not longer than twenty-five hundred words the meaning of the term so that the ordinary lay reader could understand it. The essays, 245 in number, were submitted under pseudonyms, in accordance with the rules drawn up by the Editor of the Scientific American, and were judged by Prof. Henry P. Manning, of Brown University, and Prof. S. A. Mitchell, of Columbia University.

The Five Hundred Dollar prize was awarded by the judges to Lieut.-Col. Graham Denby Fitch, Corps of Engineers, U. S. A. The prize-winning essay was published in the Scientific American of July 3rd, 1909, and three essays, which received honorable mention, made their appearance in the issues of July 10th, 17th, and 24th, 1909.

Despite the character of the subject, extraordinary interest was manifested in the contest. Competitive essays were received not only from the United States, but from Turkey, Austria, Holland, India, Australia, France, and Germany. In fact, almost every civilized country was represented. Because of this unexpected interest in the subject, it has seemed advisable to preserve in permanent form a few of the essays which were submitted. Accordingly Prof. Henry P. Manning has chosen from the essays those which lend themselves best for the purpose of a popular book on the Fourth Dimension; in other words, those which present the subject from as many different points of view as possible. With the exception of the prize-winning and honorably mentioned essays, no attempt has been made to arrange the essays in the order of merit.

The publishers trust that the brief expositions of the Fourth Dimension here offered may serve to popularize a topic which has hitherto been unfortunately classed with such geometrical absurdities as the squaring of a circle and the trisection of an angle.

CONTENTS

INTRODUCTION

BY HENRY P. MANNING, ASSOCIATE PROFESSOR OF
MATHEMATICS IN BROWN UNIVERSITY.

I.

THE geometry studied in the schools is divided into two parts, Plane Geometry, or Geometry of Two Dimensions, and Solid Geometry, or Geometry of Three Dimensions, and the study of these geometries suggests an extension to geometry of four or more dimensions. In a plane, for example, one line may be perpendicular to another, and the position of any point can be determined by starting from a known point and measuring in two given perpendicular directions. In Solid Geometry there may be three mutually perpendicular lines, and the position of any point can be determined by starting from a known point and measuring in three given perpendicular directions. Thus the question arises: Why may there not be a geometry with four mutually perpendicular lines, in which the position of a point is determined by measuring in four perpendicular directions? Again, the area of a rectangle is expressed as the product of its base and altitude, and in Plane Geometry the things that are studied are made up of straight or curved lines or are bounded by such lines. The volume of a rectangular solid is expressed as the product of its three dimensions, and the things that are studied in Solid Geometry are mostly made up of flat or curved surfaces or are bounded by such surfaces. Why then may there not be rectangular figures of four dimensions and a study of things which we may call flat or curved spaces?

The Geometry of Three Dimensions is more extensive than Plane Geometry, yet nearly everything in it is more or less analogous to something in the plane; and so the Geometry of Four Dimensions would be still more extensive, yet related to the three-dimensional geometry as the three-dimensional geometry is related to the two-dimensional, so that it would seem almost possible to tell at once what the details of such a geometry would be.

These suggestions come more readily when the real subject matter of geometry and the nature of geometrical reasoning are understood. Geometry does not deal with material things like a string or sheet of paper, but with abstract lines and surfaces. Nor does geometry deal with actual facts. It only shows what would be true if certain other things were true. We apply some statement of geometry to a string or to a sheet of paper whenever the conditions of the statement seem to be fulfilled, and the correctness of the result depends upon whether the conditions are fulfilled.

Even the axioms of geometry, formerly regarded as self-evident truths, are now understood to be merely hypotheses. The mathematician does not say that the axioms are true. He develops a set of propositions that follow necessarily from the axioms and are involved in the axioms themselves, but the axioms themselves he can change, and by taking different sets of axioms he can build up different geometries, each geometry mathematically true, true in that it is a set of statements (theorems) necessarily involved in the set of axioms upon which it is built. It is necessary.

that the axioms chosen for a geometry shall be consistent; they must not contradict one another. They ought also to be independent; no statement should be taken as an axiom if it necessarily follows from the other axioms. Finally, the set of axioms should be complete, so that the geometry is completely determined without requiring additional axioms.

We choose, then, one of these geometries and apply it to our lives. We choose that geometry whose axioms and resulting theorems seem best to express the conditions of our existence, but this choice is not a part of mathematical reasoning; it is a matter of experiment and of experience.

Finally, the mathematician may go still further and leave undefined the subject matter of his geometry. He takes certain elements, calling them points and lines, and certain relations which he calls relations of position and magnitude. Without defining the elements or the relations he assumes that the elements have these relations. The statements that the elements have the relations are his axioms. From the axioms he derives other relations which necessarily follow. The statements of these relations are his theorems.

This is abstract geometry.* The terms used are meaningless, whether they are the words *point, line, intersect, etc.,* borrowed from the ordinary geometry, or new words invented for the purpose. It is easier, of course, to assign meanings to the terms at the beginning and give to the geometry a concrete form as it develops, especially if the concrete form is not too difficult for us to picture in our minds, but it is possible to construct the geometry abstractly and then to apply it by giving concrete meanings to its terms. By

---

* This theory of abstract geometry is referred to in Essay II., p. 58.

changing the meanings of the terms we can give to the same geometry more than one interpretation even when the geometry is first constructed in concrete form.

When the student gets this view of geometry fixed in his mind he is more ready to entertain the notion of a geometry of four or more dimensions. He sees no difficulty in assuming a set of axioms which includes the hypothesis that there are points outside of a given space of three dimensions when *points* and *space* are themselves words without meaning. The difficulty which he meets in contemplating such a geometry or any geometry comes when he attempts to apply it to our existence or to some imagined existence where its application seems to contradict or to transcend our experience.

We have said that the same geometry can have more than one interpretation. Thus we shall see presently that a certain two-dimensional geometry may be interpreted as spherical geometry if we make the term *straight line* mean great circle. With a proper definition of *length* or *distance* our ordinary geometry may be interpreted as a geometry in which the circles through a certain fixed point are taken for *straight lines*. And so we might give other illustrations. Now the abstract geometry of four dimensions may be realized as a concrete geometry by letting the word *point* mean straight line in our space. It takes four numbers to determine the position of a straight line, and all the relations of the Geometry of Four Dimensions are represented by relations of these lines and by figures formed of them.*

But these interpretations seem far-fetched, and the

* See Essay XIII., p. 159; see also C. J. Keyser, "Mathematical Emancipations," The Monist, 1906.

abstract geometry itself is of interest chiefly to those few even among mathematicians who have made the theories of geometry their special study. The geometry of straight lines in space, for example, is of interest and value in itself, but that which especially interests us now is the interpretation of Geometry of Four Dimensions in its most natural way, where *points* mean points and *straight lines* mean straight lines and the relations considered are the same as those which we have in applying two-dimensional and three-dimensional geometry to our actual existence. Even when the mathematician makes use of this geometry in the study of some other branch of mathematics it is in this natural interpretation that he wants it.

The most notable of the geometries developed from different sets of axioms are two, commonly called non-Euclidean geometries. These geometries are quite fully explained in the second essay of this collection.* Neither Lobachevsky nor Bolyai thought of geometry in the abstract way that we have indicated, but the Hyperbolic Geometry, which they discovered, was one which would seem to fit very well with our experience if we confined our attention to a small portion of a plane or to a small region of space. The same is true of the Elliptic Geometry. We cannot even say that the geometry of our space is Euclidean and not one of these two. Now the non-Euclidean geometries of two dimensions can be applied to certain curved surfaces in ordinary space (the space of Euclidean Geometry) if we interpret the term *straight line* to mean geodesic or straightest line. Some writers have taken this as an explanation of the non-Euclidean

---

* See also " Non-Euclidean Geometry," by Henry P. Manning. Ginn & Company.

geometry and supposed that the plane of this geometry is not a plane and that the straight line is not a straight line.

In the same way that we have curved surfaces in ordinary space to which we can apply the non-Euclidean geometries of two dimensions, so in space of four dimensions we have curved spaces or hypersurfaces to which we can apply the non-Euclidean geometries of three dimensions, and some have taken this fact as completing the explanation of these geometries, erroneously supposing that they assume our space to be a curved space in space of four dimensions. Some have even thought that the Geometry of Four Dimensions was invented for the purpose of explaining the non-Euclidean geometries. The non-Euclidean geometries do not themselves assume that space is curved, nor do the non-Euclidean geometries of two and three dimensions make any assumption in regard to a fourth dimension. In fact, we may suppose that space of four dimensions, if there is such a space, is itself non-Euclidean, elliptic or hyperbolic as the case may be, and that our space is a three-dimensional space of the same kind without any curvature whatever. The notion of a geometry of four dimensions does not owe its origin to the non-Euclidean geometries. We have the same breaking away from tradition in both and both grow out of modern theories of the general nature of geometry, but the geometries of higher dimensions owe their origin to a natural extension from two and three dimensions and the mathematician has other uses for them equally as important as is their relation to the non-Euclidean geometries.

The notion of geometries of higher dimensions takes on its chief importance in Mathematics from the parallelism between Algebra and Geometry. Algebra had been used to some extent in the proofs of theorems which involve proportion and other relations of magnitude, but the study of Algebra and Geometry together was first systemized in Analytic Geometry and became thereafter the basis of a great part of Mathematics. Now certain forms of Algebra correspond to Plane Geometry and certain other forms to Solid Geometry. Besides these there are also what might be called one-dimensional forms, and no difficulty is found in realizing the corresponding geometry as a geometry of points on a line, although this geometry would hardly have attracted attention had it not been for the needs of Algebra.

This combination of Algebra and Geometry, which appears at first sight to serve chiefly as an aid to Geometry, turns out to be of greater service to Algebra. This happens in two ways. The language of Geometry furnishes a number of convenient terms for things which would otherwise have to be awkwardly described, and the visual conceptions of Geometry applied to the forms of Algebra make them seem less abstract and easier to understand. We have these advantages for the forms of Algebra which correspond to geometries of one, two, and three dimensions. Yet there is no reason in Algebra for the distinction between these forms and other forms, and when we have become accustomed to apply geometrical terms in Algebra we begin to use them in connection with all algebraic forms and thus to secure the first of the two advantages mentioned as derived from the combination of Algebra and Geometry.*

But it is from the visual conceptions of Geometry

---

* See Essays V., IX., and XIV.

that the mathematician gets his chief assistance when he applies Geometry to Algebra, and since the geometries of higher dimensions are necessary to the complete parallelism of the two, he seeks to acquire these conceptions here also by trying to imagine our existence in a space to which these geometries apply. This is especially true of the Four-Dimensional Geometry to which correspond some of the most important forms of Algebra.

We find, then, two ways in which the geometry of four or more dimensions is of importance to the mathematician. The notion of such a geometry as a logical system of theorems involved in a set of axioms is important to the student of abstract geometry, and the conception of space to which these geometries apply is of great assistance in the application of geometry to other mathematics. No one can consider himself completely equipped as a mathematician without some knowledge of the geometries of higher dimensions.

II.

The notion of geometries of n dimensions began to suggest itself to mathematicians about the middle of the last century. Cayley, Grassmann, Riemann, Clifford, and some others introduced it into their mathematical investigations. Then from time to time different mathematicians took it up in different ways. Thus the first volume of the American Journal of Mathematics begins with an article in which Professor Newcomb shows that a sphere may be turned inside out in space of four dimensions without tearing, and in the third volume of the same journal Professor Stringham has given us a full account of the regular figures in space of four dimensions corresponding to the regular polyhedrons of our three-dimensional space. Others have written on the theory of rotations and on the intersections and projections of different figures. The great Italian geometer Veronese has an extensive work on Geometry of n Dimensions with theorems and proofs like those of the text-books studied in our schools. In the last few years there have been many articles in the popular magazines, and some books have been published to explain more particularly what the fourth dimension is.* The fourth dimension is the first of the higher dimensions and in this book it alone is considered.

Geometry of Four Dimensions is not only of importance to the mathematician, but it is also of interest in certain other lines of study. Thus it involves questions of space which concern the philosopher; efforts to understand it call into exercise our space perceptions and so attract the attention of the psychologist; and attempts to utilize the theories of hyperspace in the explanation of physical and other phenomena serve to bring the subject under the notice of those working in other branches of science. Moreover, the many curious forms and relations that appear in its development excite popular interest; for example, the relation of symmetrical forms as one of position only, a form being changeable into its symmetrical by mere rotation; the plane as an axis of rotation, and the possibility that two complete planes may have only a point in common; the possibility that a flexible sphere may be turned inside out without tearing, that an object may be passed out of a closed box or room without penetrating the walls, that a knot in a cord may be

* Some references are given at the end of this introduction.

untied without moving the ends of the cord, and that the links of a chain may be separated unbroken.

These curious features of space of four dimensions, while exciting our interest, baffle us in our study. Not only the possibility of such things but the facts themselves seem beyond our comprehension. In Plane and Solid Geometry we can draw figures and construct models; we are constantly seeing the things themselves and therefore, even when they are complicated, we can readily picture them in our minds. Geometry of Four Dimensions, however, in its ordinary application, deals with things which no one has known in experience or can imagine. Its very words seem to have no meaning. This is especially true at first, and any facility in perceiving the relations of these words, if acquired at all, must come slowly and of itself. In our efforts to understand the subject we naturally desire a perception of these things at the beginning. All that we should try to do, however, is to remember the various relations and to become familiar with them. In time they may perhaps acquire some of the vividness of the conceptions of three dimensional geometry. If we expect too much when we begin this study we shall be disappointed and discouraged. If we understand at the outset how little we should expect, we shall be in an attitude toward the subject that will be most conducive to success in its mastery.

It follows that we shall not find this subject an easy one to understand. It is something that we have to read a little at a time, to read repeatedly and to think over. We have to look at it from different points of view and to examine different ways of expressing it. Thus there are distinct advantages in having the subject presented in several short essays by different writers. There are advantages in the repetition, in the different points of view, and in having brief independent chapters that can be taken up and studied each by itself.

The essays in this book are all non-mathematical or popular in their treatment. It will assist us, therefore, if we understand the limitations of this form of presentation. From a comparison of the lower dimensional geometries we derive analogies for the Geometry of Four Dimensions and the analogies are so complete that the subject can be very fully explained in a non-mathematical way. The analogies are a guide, even to the mathematician, but the geometry does not depend on these analogies. As a system of theorems and proofs it is built up from its axioms by a process of logical reasoning just as the lower geometries are built up. If we wish to be convinced of the consistency of this geometry, of its truth as a mathematical system, we should study it mathematically. A non-mathematical exposition should be received solely as an explanation of the geometry itself, and the reader should understand clearly that it is designed not to convince him even of the possibility of such a geometry, but only to show him what it is.

The adoption of such an attitude on the part of the reader will be a long step toward accomplishing all that can be achieved through a non-mathematical treatment of the subject. If, however, the analogies are viewed as arguments, a person of skeptical mind will be apt to suspect that there is some fatal defect beneath their plausible exterior. Even if a philosophical writer wishes to use the analogies as well as the consistency of this geometry as an argument for the actual existence of four-dimensional space, such a consideration of the subject had better be postponed by the reader until after he has become familiar with the geometry itself. As regards some of these essays it is proper to caution the reader that they seek to advocate certain views rather than merely to give a clear description of the fourth dimension.

There is another way in which the principle of analogy may be used. By imagining two-dimensional beings living in a plane and unable to perceive anything of a third dimension we get a vivid idea of our own relation to four-dimensional space. A consideration of what ought to be their attitude toward any conceptions of a space of three dimensions makes clearer what should be our attitude toward conceptions of a higher space. This point of view is made more interesting by presentation in story form of a picture of life as it might be supposed to exist in a two-dimensional world. It is not necessary for such a presentation to go into all the details of the two-dimensional existence. A too minute description of such an existence would overburden the narrative with tedious explanations that would cause us to lose sight of its main purpose. But a story written so as to bring out skillfully a few of these relations does very much to help us in understanding what should be our attitude toward the higher geometry.*

The Geometry of Four Dimensions based on a suit-

* Such a book has recently been written by C. H. Hinton: "An Episode of Flatland." But much better is a little book by E. A. Abbot called "Flatland." There the interest rests entirely on the relations of space which this book is intended to explain, and we never for a moment lose sight of these relations. In Hinton's book the interest rests largely on the personalities and fortunes of the characters, and our attention is called away from the geometrical circumstances of their lives. Moreover, his circle-world is more unreal than the world of "Flatland," although, perhaps, more closely analogous to our earth as it exists in space of three dimensions.

able set of axioms and applied in the ordinary way to points, lines, etc., forms a definite system. But there is much that is arbitrary when we come to clothe our ideas in physical form and undertake to present a material world either of two or four dimensions, filled with two-dimensional or four-dimensional matter. Even to the physicist matter is a mystery and we can develop different theories of it very much as we build up geometries from different sets of axioms. Some writers of these essays have made quite unwarranted statements as to what *must* be the nature of matter. We cannot say that we have perceived all the properties of matter as it exists, and we cannot call it absurd to put matter with other properties into an imagined space. Thus in order to throw light upon our relations to a supposed space of four dimensions we might suppose the existence of two-dimensional beings even if such an existence were impossible, just as we might imagine the moon inhabited by intelligent beings in order to give a more vivid description of the appearance of the moon's surface by describing what they would see. We do not know, indeed, but that the moon is inhabited by beings with bodies adapted to their environment, capable in some way of life, growth, and motion, without air or water.

In thus supposing the existence of two-dimensional beings it would be interesting in itself to see how far we can go in these details. Thus we may suppose that what we call two-dimensional matter is really three-dimensional, and that the two-dimensional beings are really three-dimensional, either with a slight thickness in the third dimension, or at least with a thickness which the beings themselves are unable to recognize. But we may also suppose them all to be really two-dimensional, and we can try to carry out the details of such an existence. It may be that a particle of matter is only a bundle of forces, attractive and repellent, and there is no difficulty in thinking of such forces lying entirely in one plane. A two-dimensional being, meeting some object, might find it, that is, its contour, hard or soft; light waves traveling in this plane might be reflected by objects, the edges of objects, and produce images on the retina line of the two-dimensional being's eye; and sound waves might strike a vibrating chord in the two-dimensional being's hearing cavity. Objects could be fastened together, either by adhesion or by one object grasping another. Mechanical contrivances and organic bodies would be of comparatively simple structure, if, as in our world, two entirely separate objects had no appreciable influence on each other. No object could have an opening through it like a hole and there would be nothing to correspond to our pipes. If a house had more than one outside door open, or if its windows were opened, it would be divided into separate parts. It would seem as though such simple forms and structures as would make up a two-dimensional existence would allow but little mental development to the inhabitants, but we find nothing impossible in the supposed structure of such a world.

When we come to consider a two-dimensional space and a three-dimensional space together, the two-dimensional space lying within the three-dimensional, we have a considerable choice as to the nature of matter in these spaces, and any apparent difficulties may be ignored without affecting the usefulness of these suppositions for purposes of analogy. We may, however, be interested in the question for its own sake and try to see how far we can carry the details of such a combination of spaces. Let us suppose that the two-dimensional matter of the plane inhabited by our two-dimensional beings has the property of reflecting in some measure light that comes from outside of this plane, so that three-dimensional beings are able to see the two-dimensional matter. They can see, then, the insides of the two-dimensional beings and the insides of their houses and within all their closed compartments. If also they are able to take objects out of this plane and put them back wherever they please, they can take them out of the closed compartments.

A study of the laws of four-dimensional matter, Four-dimensional Physics, would be very interesting, but we can give some idea of the various forms which occur, and the possible motions of things, without going too carefully into the theory or using the terms of science with great exactness. Our object is to give some idea, something as near a picture as we can, of the space of four dimensions, and we shall impose limitations upon the beings which we describe, or remove limitations, according to the course which seems best adapted to our object.

We observe the forms and positions of objects very largely by sight. Now the organs of sight of a being confined to some particular space may be supposed suited to the dimensions of his space. The picture formed in the retina of our eye is two-dimensional, the retina is a surface. A two-dimensional being, unable to perceive anything outside of his plane would have a one-dimensional retina, or at least his picture of an object in his world would be a mere line, different pictures being distinguished by the lengths, colors, and shading of these lines. The retina of a four-dimensional being would be three-dimensional if he is to receive separate impressions from all the rays of light within a given angle of vision. In fact, the boundary of an opaque object, the part which alone he can see, is three-dimensional as is always the boundary of objects in space of four dimensions.

It is not easy for us to imagine such pictures, and so we can attempt to get an impression of the shapes of objects by supposing that a three-dimensional being, a person like ourselves, could pass through a series of parallel three-spaces (three-dimensional spaces) and in each three-space examine that portion of the object which lies in this space, that section of the object. This is just as we might suppose a two-dimensional being able to pass through a series of planes and in each plane to see the section of an object made by that plane. The section which we should see of a four-dimensional object would be a solid whose surface forms a part of the three-dimensional boundary of the object. This way of studying four-dimensional objects is discussed quite fully in Essay VII. (See also Essay V, page 85.)

There is another somewhat similar way of studying an object that we may find quite useful. We can imagine ourselves turning from one three-space into another perpendicular three-space. That is, by discarding one of the directions in our space we can suppose that we take into view the fourth direction, which goes away from our space, and so get its relation to two of our directions. We shall describe the section of an object made by any three-space as what we can see in that three-space. We shall do this particularly with reference to the different sections of an object obtained at any point by taking different perpendicular three-spaces.

One of the first things, for example, that we consider in studying Geometry of Four Dimensions is the
line perpendicular to a three-space; such is the line
which goes out from a point in our space in a new
fourth direction perpendicular to all the lines of our
space through that point.* If we can let go of one of
the dimensions of our space, keeping only that part
which lies in a certain plane, and take into view the
new fourth dimension, we shall see a plane and a line
going out from it, perpendicular to all the lines of it,
something with which we are perfectly familiar.

As another example consider two absolutely perpendicular planes. If we take a plane through a point
O and the line which is perpendicular to the plane at O,
all in our space, and then take the line through O in
the fourth direction perpendicular to all the lines
through O in our space, we shall have a plane through
O and two lines both perpendicular to the plane and
perpendicular to each other. These two lines themselves determine a plane every line of which through O
is perpendicular to the first plane. The two planes are
said to be *absolutely perpendicular*. (See Essay I, page
45, where the expression *completely perpendicular* is
used.) The most that we could see in any three-space
of two absolutely perpendicular planes would be one
of the planes and a single line of the other plane, a line
passing through O perpendicular to the plane that we
see. The other plane cuts through the space along
this line. These planes meet only at the point O.
Indeed, two planes which do not lie entirely in one

---

* A point starting from the center of a sphere in our space and moving off
on a line perpendicular to our space will not approach any portion of the
surface of the sphere, but will move away at the same rate from all points of
this surface. This is the way an object can pass out of a closed room or box
without penetrating the walls, as stated in many of the essays.

three-space can never have more than a point in common, and when two planes have just a point in common the most that we could see in any three-space would be one of the planes and a single line of the other.

If two planes are absolutely perpendicular to a third at two points O and O' they lie in a single three-space. In this three-space we should see them completely, and only a single line of the third plane. The line passes through O and O' and we see it as perpendicular to the two planes. On the other hand, in a three-space containing the third plane we can see all of it but only a single line of each of the two planes absolutely perpendicular to it.

III.

We proceed to give some further account of the Four-dimensional Geometry. We do not intend to repeat what is given in the essays except so far as may be necessary in order to correct possible erroneous impressions, or to amplify certain points. It may be that the reader will find it better to read some of the essays before going on with this Introduction.*

When two planes are absolutely perpendicular at a point O, any point in one can go completely around O and around the other plane keeping all the time at the same distance from O and from the other plane. Thus we can go around a plane in space of four dimensions just as in our space we can go around a line. A two-dimensional being cannot go around a line in his plane; it divides the plane completely. And so we cannot go around a plane in our space for it divides

---

*There is quite a diversity in the use of terms in Geometry of Four Dimensions. Most of the terms used in this book, however, are defined when used or are readily understood.

our space completely. But in space of four dimensions a plane, though having two dimensions, lacks two, and in these we can go around the plane keeping all the time at a given distance from one particular point of it. If we can discard one of the dimensions of the plane, taking from the plane only a line, and put ourselves into a three-space that contains the absolutely perpendicular plane, we shall find that the path of the motion is all in view, appearing to us now as a path going around a line.

A plane can rotate on itself around one of its points. If two planes are absolutely perpendicular at a point O, one of them, rotating on itself in this way, remains absolutely perpendicular to the other. We may speak of the plane as rotating about the fixed plane as axis plane. At each point of a fixed plane is an absolutely perpendicular plane and these absolutely perpendicular planes may all rotate together about the fixed plane. This is the same as when we have in our three-space a fixed line and at each point a plane perpendicular to the line. Thus we think of objects in our space or of a portion of space as rotating about a fixed axis line; and in the same way we can think of objects in four-space or of a portion of four-space as rotating about a fixed plane as axis plane. In this rotation the parts are not distorted; they retain their form rigidly and need not be flexible.

We may also use a curved surface as axis of a rotation if we allow for a slight amount of distortion. We will use the term *material surface* for a substance having two dimensions of considerable extent and two dimensions very small, just as we may say in our space that a piece of cloth has two dimensions of considerable size and one dimension very small, or that a string has one principal dimension and two dimensions very small. If we have such a material surface that is flexible, we may rotate it, each portion on itself, so that two opposite sides of it shall exchange places. A material surface, like a piece of cloth with a slight thickness in the fourth dimension, will have surfaces all around it. We may say that a turning of such a substance on itself through 180 deg. brings the same two sides back into our space, each on the side originally occupied by the other. The different parts of the surface do not interfere with one another in this process, and so it may take place whether the surface is open, any piece of a material surface, or completely closed like a hollow rubber ball. In our space a rubber band may be twisted on itself so as to be turned inside out. This corresponds exactly to the turning of a sphere inside out in space of four dimensions.

The relation of symmetrical figures is referred to in several of these essays but not always quite correctly. Symmetrical figures can best be understood by considering positions of symmetry with respect to a point, line, or plane.

Figures in a plane symmetrical with respect to a point are equal, for one can be turned about the point to the position of the other. Figures in a plane symmetrical with respect to a line, however, cannot be made to coincide without turning one of them over, turning it through space. Such figures would be regarded by two-dimensional beings as truly symmetrical, with corresponding parts equal, but arranged in opposite orders, so that it would never be possible to make them coincide.

Figures in space of three dimensions symmetrical

with respect to a line can be made to coincide by turning one of them about the line. On the other hand, figures symmetrical with respect to a point and figures symmetrical with respect to a plane, unless they are actually plane figures, are truly symmetrical and can never be made to coincide by a motion in space. Figures symmetrical with respect to a plane can be made to be symmetrical with respect to a point, and figures symmetrical with respect to a point can be made to be symmetrical with respect to a plane. Suppose, for example, two figures are symmetrical with respect to a plane. We connect them by a rod perpendicular to the plane and join pairs of corresponding points by lines, say elastic cords. Then if we turn one of them half-way around on the rod as axis the elastic cords will all cross one another at the point where the axis rod meets the original plane of symmetry, and they will become symmetrical with respect to this point.

Now in space of four dimensions figures may be symmetrical with respect to a point, a line, a plane, or a three-space. Figures symmetrical with respect to a point may be made to be symmetrical with respect to a plane and *vice-versa*, and figures symmetrical with respect to a line may be made to be symmetrical with respect to a three-space and *vice-versa*. Figures symmetrical with respect to a three-space are truly symmetrical and can never be made to coincide by any motion in four-dimensional space. They may be said to have their parts arranged in opposite orders. But figures symmetrical with respect to a plane may be made to coincide by rotating one of them about the plane as axis plane through a rotation of 180 degrees, and this is true whether they are four-dimensional figures or three-dimensional figures. Thus to a four-dimensional being things which we call symmetrical do not differ at all except in position.

This is a very striking fact. A right glove turned over through space of four dimensions becomes a left glove, a right shoe becomes a left shoe. A right-handed man becomes a left-handed man. He does not use a different hand after the operation, but the hand which he uses now appears to everybody else as his left hand. In fact, his point of view is turned around, so that to him everybody else appears to be changed. Letters appear to him to be turned backward like printer's type, the hands of a clock go backward, the world becomes to him a looking-glass world.

There is a distinction not understood by some of these writers between turning an object over and turning it inside out. A right glove turned inside out in our space becomes a left glove and a right glove turned over in space of four dimensions becomes a left glove, but when the glove is turned over it is not turned inside out.* On the other hand, a right glove may be turned inside out in space of four dimensions in the same way that a closed rubber ball may be turned inside out. This process has been described in a preceding paragraph. The fingers and thumb do not come out through the wrist, but every part by itself in its own place is turned over with only a little possible stretching and a very slight changing of the positions of the different particles of matter which go to make up the glove. In this process, however, the glove does not become a left glove, but remains a right glove. We can get the analogy by supposing that we have in a plane a nearly closed figure. This can be turned into

* Even Schubert makes this mistake in his article, " The Fourth Dimension," The Monist, Vol. III., page 429.

its symmetrical form by opening it out straight and bending it over the other way so that it is turned inside out. This process takes place entirely in the plane and can be performed by a two-dimensional being. The figure may also be changed into its symmetrical form by being turned over, but in this process it is not turned inside out at all. On the other hand, if it is sufficiently flexible, it may be turned inside out by twisting each part upon itself through 180 degrees, and in this process it is not changed into its symmetrical form.

A hypersolid, that is, a portion of four-dimensional space, may be separated into two parts by a three-space. Thus a section, cutting a hypersolid into two parts, will be three-dimensional. A plane cannot separate two parts of a hypersolid any more than a line can separate two parts of a solid in our space. We may make a line go through a solid, cutting out a hole. This may happen also to a hypersolid. A rod or material line, having one principal dimension and the other three very small, will pierce a hypersolid and make a hole through it. But we may also pierce a hypersolid with a flat plate, something having two principal dimensions and two dimensions very small. The plate passing through the hypersolid could extend indefinitely in its two principal dimensions but the hypersolid would not fall apart. Thus we have two kinds of holes in space of four dimensions, one-dimensional holes and two-dimensional holes.

A one-dimensional hole may pass through an object in a direction away from our space and the object will appear to us entirely closed but hollow like a hollow sphere. A rod or cord may be passed through such a hole and held by it in position, but a rod or cord passed through a two-dimensional hole will slip away at once even if we hold its ends. A rod bent around so that its ends can be welded together becomes a ring. The hole through a ring is two-dimensional. Two rings fall apart, but a ring and a hollow sphere may be linked together. Thus we may form a chain of alternate rings and hollow spheres. In an ordinary knot one end of a cord is passed through a ring formed of the cord itself and slips away at once in space of four dimensions.*

A wheel of four-dimensional matter, in two dimensions of the shape of a circle and in the other two dimensions very small, would have for axis a flat plate instead of a rod. This axial plate could extend indefinitely in all the directions of its plane† without any interference with the wheel. The wheel can slip all around over the axial plate unless held to some position on it, just as with us a wheel may slip along on its axis unless held to some position on it. We may suppose that in a three-space we can see the axial plate and a pair of opposite radii (spokes) of the wheel, appearing to us entirely separate; in this way we can see a two-dimensional hole. Or we can see the entire wheel with a hole through it and an axial rod, cut from the axial plate by our three-space.

We can fasten the wheel rigidly to the axial plate so

---

* Some of the writers speak of a loop or "two-dimensional knot " as analogous to an ordinary knot made with a string in our space. This analogy seems to have been used by Zöllner, but there is the objection that the loop is not two-dimensional if one part of the string passes over the other part, however closely they may be pressed together. A better analogy would be obtained by fastening a string at one end to a small object and winding it around this object. In the plane this would be possible only by carrying the free end of the string completely around, but we could do it in space of three dimensions by lifting a part of the string over the object without moving the free end away from its position.

† Sometimes we shall speak roughly of the plane of the wheel or the plane of the plate just as we might in our three-dimensional world.

that it will turn with the wheel, the wheel turning in its plane and the plate turning on itself. We may put more than one wheel on an axial plate, putting different wheels at different points on the plate wherever we please. If these wheels are all fastened rigidly to the axial plate we turn them all by turning one. Thus we have a method of constructing machinery in space of four dimensions.

The axial plate may itself be a wheel. We may fasten two wheels together at their centers making them absolutely perpendicular to each other. Such a figure can revolve in two ways, the plane of one wheel being the axis plane of the rotation and the plane of the other wheel the rotation plane.

A wheel may be doubly circular so that a plane absolutely perpendicular to the wheel cuts it in a small circle just as the plane of the wheel itself cuts it in a large circle. Such a wheel, then, may turn in two ways and in either kind of rotation it rotates completely on itself without passing through any new portions of its four-dimensional space. (See below, page 38.)

We might have a spherical wheel; something in three dimensions of the shape of a sphere and its fourth dimension very small. Such a wheel with a one-dimensional hole through it may turn on an axial rod, but its motion is not confined to a definite direction of rotation as is the case with the flat wheel turning in its plane. For machinery requiring definite rotations we should use flat wheels with axial plates.* A spherical wheel

---

* Hinton speaks of the "four-dimensional being's shaft, a disk rotating around its central plane," and of the spherical wheel, "the four-dimensional wheel." ("The Fourth Dimension," pages 61 and 71-3.) By associating these he leaves an impression that the axis of his wheel is his disk, whereas his wheel has a one-dimensional axis and is not the kind of wheel to be used with his four-dimensional shafting.

may be used for vehicles. If four dimensional beings lived on a four-dimensional earth; that is, alongside of its three-dimensional boundary, a vehicle with four or more wheels of either kind could be used in traveling over this earth. With a flat wheel he could travel only in a straight line without friction between the wheel and the earth; with a spherical wheel he could travel in any direction in a plane without such friction, but would meet with a slight friction in turning from one plane to another.

A vehicle would require at least four wheels to be in equilibrium, and these must have at least two axes. Even with flat wheels and axial plates it is necessary to have at least two of these plates. Anything to remain in equilibrium must have at least four points of support, not all in one plane.

It is difficult to comprehend how the boundaries of hypersolids, that is, of portions of four-dimensional space, are three-dimensional. It is evident that analogy requires this, but it is not easy to understand how each point within a solid can be all that in its place separates the two portions into which the three-space of the solid divides four-space. At any point in the three-dimensional boundary of the hypersolid we can start and go in three mutually perpendicular directions within this boundary—in as many directions as we have altogether in our three-dimensional space. We may have to trace curved paths if the boundary of the hypersolid is curved, but the paths start out in three mutually perpendicular directions just as in our space.

We can cut open a hypersolid bounded by polyhedrons so as to spread them out in a single three-space. Reversing this process, we can form the boundary of a hypersolid by putting together suitable solids in a three-space, say in our space, and then turning them on the faces which join them until they are all brought together. The solids are not distorted in any way nor separated. Thus if we take a cube, place six equal cubes on its six faces and one extra cube on one of the six (see Essay IV, Fig. 4 and context), these can all be turned and brought together to form the hypercube or tesseract which many of the essays describe. We have the analogy in the case of polyhedrons whose faces can be cut apart sufficiently to spread them out in a single plane. The analogy is so very clear that we may feel sure of the process, although the result is most puzzling.

We shall mention some of the simpler figures of four-dimensional geometry corresponding to the figures studied in our solid geometries.

Among the first to be noticed are the hyperprism and hypercylinder with parallel line elements, and the hyperpyramid and hypercone with line elements meeting at a vertex. These all have for bases polyhedrons or solids of some kind, and the element lines extend away from the three-space of the base. The hypercube is a very particular case of the hyperprism.

The simplest case of a hyperpyramid is a figure called a pentahedroid. It has for base a tetrahedron or triangular pyramid and thus it has in all five vertices. Any five points, not all in one three-space, may be regarded as the vertices of a pentahedroid. These five points, taken four at a time, give us five tetrahedrons and the pentahedroid may be taken in five different ways as a hyperpyramid. The tetrahedrons are placed together face to face, each having one face in common with each of the others. We can cut these tetrahedrons apart sufficiently to spread them out into one three-space. We then have a single tetrahedron with four others resting on its four faces. The pentahedroid is formed by turning these toward one another until they are brought completely together again. In this process none of the tetrahedrons is distorted nor are they in any way separated from one another. When brought completely together they form a single closed figure inclosing a portion of hyperspace. This is analogous to the way in which we can spread out the faces of a tetrahedron in a single plane, and, reversing the process, bring them together again and form the tetrahedron.

In general, the boundary of a hyperpyramid consists of the polyhedron base and of lateral pyramids resting on the faces of the base. The lateral pyramids are joined to one another by their lateral faces in the same way that the faces of the polyhedron base are joined by the edges.

A hyperpyramid whose base is a pyramid may be regarded in two ways as a hyperpyramid, the vertex in either case being the vertex of the pyramid base in the other case. The two pyramid bases have, then, a common polygon base and the hyperpyramid may be considered as determined by a polygon and two points not both in a three-space with the polygon. The line joining the two points may be called a line-vertex and the boundary consists of the two pyramids and a portion which may be generated by a triangle, varying it may be as to size and shape, with one side fixed, and with the opposite vertex tracing a polygon which does not lie in a three-space with the fixed side. The generating triangle may, then, be called a triangle element.

Similarly, a hypercone with a cone for base may be regarded in two ways as a hypercone and has for boundary the two cones and a portion generated by a triangle with one side fixed, the opposite vertex tracing a plane curve which does not lie in a three-space with the fixed side.

The boundary of a hyperprism consists of the two polyhedron bases and a set of lateral prisms. The lateral prisms have for bases the faces of the polyhedron bases of the hyperprism and are joined to one another by their lateral faces.

A hyperprism with prism bases has for lateral boundary two prisms and a set of parallelopipeds. Such a figure may be considered in two ways as a hyperprism, the two lateral prisms in one case being the two bases in the other case. The four prisms are joined in succession by their ends and the series of parallelopipeds are joined, each to the two next to it, by two opposite faces and to a lateral face of each of the four prisms by the remaining four faces. If the four prisms are cut away from the parallelopipeds and cut apart along one common base they can be spread out in a single three-space, and if they are right prisms they become a single right prism. The parallelopipeds may then be cut apart along one common face and spread out in like manner, forming when rectangular a single right prism (parallelopiped). These two long prisms may be placed together on any pair of faces that were originally together, one prism placed crosswise to the other, and then they may be turned from face to face all over one another. In the original figure they were wound around each other in such a way that every point in the lateral surface of one fitted upon a point in the lateral surface of the other, and they completely inclosed a portion of four-dimensional space.

If from the four prisms are taken four elements

that form a parallelogram, the set of parallelopipeds may be generated by moving this parallelogram parallel to itself, its vertices tracing the ends of the prisms. The set of four prisms may also be generated by one of the polygon bases moving parallel to itself, its vertices tracing the parallelograms which join the parallelopipeds to one another. Thus the parallelogram and the polygon play the part of generating elements, each with the other for directrix in generating a portion of the hyperprism.

In a similar way we may have a hypercylinder with cylinder bases. A part of the lateral boundary consists of two cylinders joining the ends of the cylinder bases, and the figure may be taken in two ways as a hypercylinder. Four elements that form a parallelogram may be taken from the four cylinders and the remaining part of the lateral boundary may be generated by this parallelogram moving parallel to itself, its vertices tracing the ends of the cylinders. Since the cylinders may be generated in a similar manner by a plane curve moving parallel to itself around any one of the parallelograms, we have a parallelogram and a closed plane curve, each playing the part of generating element with the other for directrix in generating one portion of the hypercylinder.

The hyperprism with prism bases and the hypercylinder with cylinder bases are, then, particular cases of a class of hypersolids which may be described as follows: Two polygons, or two closed plane curves, or a polygon and a plane curve are placed together so that they intersect but do not lie in a single three-space. Their planes will intersect only in the point where the curves intersect. One polygon or curve moves parallel to itself around the other and generates (with all of its interior points) a ring-shaped three-dimensional figure. The other polygon or curve moves in like manner around the first, generating a second ring-shaped figure. These two ring-shaped figures fit completely, and together form the boundary of a hypersolid, inclosing a portion of four-space. We may call the hypersolid a double prism, a prism-cylinder, or a double cylinder according as we have two polygons, a polygon and a curve, or two curves. When the planes of two generating polygons are absolutely perpendicular we have a right double prism, and so for the others.

If either portion of the boundary is separated from the other and cut through along one generator it may be spread out into a single three-space like our space. When the planes of the two generators are absolutely perpendicular each portion of the boundary spread out into a single three-space becomes a right prism or a right cylinder. We may in this case describe these figures in another way. To form a right double prism, for example, we take two right prisms with the altitude of each equal to the perimeter of the other. We can then bend these around each other, bring them together completely in all parts of their surfaces, and inclose a portion of four-dimensional space. In the same way we can form a right prism-cylinder or a right double cylinder, taking in one case a prism and a cylinder and in the other case two cylinders.

When cylinders of revolution are taken in this way the double cylinder formed may be called a cylinder of double revolution. This can rotate in two ways independently about two absolutely perpendicular planes, the planes of the circles formed from the axes of the two cylinders. In each of these rotations one of the axis circles rotates on itself and the other, lying in the plane which is the axis of the rotation, remains stationary.

When one of the component cylinders has a very small radius in comparison with the other, so that the second has a very small altitude, one cylinder being like a rope and the other like a wheel,* the hypersolid is what we have called a doubly circular wheel (page 31).

One more figure which we have in four-space is the hypersphere mentioned in one or two of the essays, the locus of points at a given distance from a fixed point. Sometimes the term hypersphere is used to denote the hypersolid, the portion of four-space inclosed by this locus, which is then called the boundary or hypersurface of the hypersphere. The hypersphere (that is the boundary) is three-dimensional, and in it we have three-dimensional Elliptic Non-Euclidean Geometry just as the ordinary spherical geometry is two-dimensional Elliptic Non-Euclidean Geometry.

We will state some of the rules of mensuration for Geometry of Four Dimensions. In the case of hypersolids there are rules for computing the volume of the boundary or of portions of the boundary, and for computing the hypervolume, that is, the magnitude of that portion of four-space inclosed. These rules may be derived for the most part as the corresponding rules for area and volume are derived in the ordinary geometry, or they may be obtained by the methods of the Calculus. They all apply to regular figures and most of

---

* Here we mean a three-dimensional rope such as we are accustomed to see in our ordinary space. All the prisms and cylinders which we have just been discussing are three-dimensional, and go to make up the boundaries of hypersolids. On the other hand, the axial plates and rods, as well as the flat wheels and spherical wheels, spoken of on pages 30-32 are four-dimensional, having some thickness in all four dimensions.

them can be extended to certain other classes of figures, but these cases need not be taken up here.

Hyperprism and hypercylinder:

    Lateral volume = Area of the surface of the base multiplied by altitude.

    Hypervolume = Volume of the base multiplied by altitude.

Hyperpyramid and hypercone:

    Lateral volume = Area of the surface of the base multiplied by 1/3 of altitude.

    Hypervolume = Volume of the base multiplied by 1/4 of altitude.

Double prism, prism-cylinder, and double cylinder:

    Volume of one portion of the boundary = Area of generating polygon or curve multiplied by the perimeter of the directrix.

The total volume of the boundary is the sum of two such products. We may say that the total volume is the sum of the two products formed by multiplying the areas of the generating polygons or curves, each by the perimeter of the other polygon or curve.

Hypervolume = Product of the areas of the two generating polygons or curves.

For the cylinder of double revolution of radii R and R' we have the formulæ,

    Volume $= 2\,\pi^2 RR'\,(R + R')$

    Hypervolume $= \pi^2 R^2 R'^2$

Hypersphere:

    Volume (of the boundary) $= 2\,\pi^2 R^3$

    Hypervolume (inclosed) $= \tfrac{1}{2}\,\pi^2 R^4$

A cylinder of double revolution circumscribed to a hypersphere, its radii equal to the radius of the hypersphere, will have double the volume of the hypersphere and double the hypervolume of the hypersphere.

IV.

The question of the existence of space of four dimensions is one which we cannot escape. It may be well to remind the reader that this is not a mathematical question, though the most interesting of all. The possibility that we are a part of a four-dimensional space with physical limitations which confine us to a three-dimensional space, and with limitations of our senses which prevent us from perceiving anything outside of this space—this possibility excites the interest of all who are inclined to abstract speculation. Attempts may be made to discover physical proofs of such a space, to build up theories on its basis that will explain discoveries of modern Physics as yet but little understood, or by it to account for various mysterious phenomena. Most of us are satisfied that no real proofs of the existence of space of four dimensions will be found along these lines. Even a workable hypothesis based on the existence of four-dimensional space, though it might serve temporarily better than any other hypothesis, would hardly justify a belief in this existence. But we do say that the existence of space of four dimensions can never be disproved by showing that it is absurd or inconsistent; for such is not the case. Nor, on the other hand, will the most elaborate development of the analogies of different kinds ever prove that it does exist.

The following articles and books treat in a non-mathematical way of the fourth dimension or other modern ideas of geometry discussed in this book:

E. A. Abbott, Flatland; Little, Brown & Co.

H. A. Bruce, The Riddle of the Fourth Dimension; Scientific American Supplement, vol. 66, p. 146.

T. P. Hall, The Possibility of a Realization of Four-fold Space; Science, May 13, 1892.

C. H. Hinton, The Fourth Dimension; Harper's Magazine, July, 1904.

C. H. Hinton, published by Swan, Sonnenschein & Co.:
Scientific Romances,
A New Era of Thought,
The Fourth Dimension,
An Episode of Flatland.

C. J. Keyser, Mathematical Emancipations; The Monist, vol. 16, 1906, p. 65.

Simon Newcomb, Modern Mathematical Thought; Bulletin of the New York Mathematical Society, vol. 3, January, 1894, p. 104.

Simon Newcomb, The Philosophy of Hyperspace; Bulletin of the American Mathematical Society, second series, vol. 4, February, 1898, p. 187.

Simon Newcomb, The Fairyland of Geometry; Harper's Magazine, January, 1902.

"S.," Four-Dimensional Space; Letter to the Editor, Nature, vol. 31, March 26, 1885, p. 481.

Hermann Schubert, The Fourth Dimension; The Monist, vol. 3, April, 1903, p. 402. Reprinted in Mathematical Essays and Recreations; Open Court Publishing Company.

J. F. Springer, The Fourth Dimension Simply Explained; Scientific American, vol. 98, p. 202.

O. Veblen, The Foundations of Geometry; Popular Science Monthly, vol. 68, p. 21.

A very good treatment of the subject in German is:

Dr. Carl Cranz, Gemeinverständliches über die sogenannte vierte Dimension; Sammlung von Virchow und Wattenbach, Hamburg, 1890.

I.

AN ELUCIDATION OF THE FOURTH DIMENSION.

THE PRIZE-WINNING ESSAY.

BY "ESSAYONS" (LIEUT.-COL. GRAHAM DENBY FITCH, CORPS OF ENGINEERS, U. S. A.)

It is impossible to form a mental picture of the fourth dimension. Nevertheless, it is not an absurdity, but a useful mathematical concept with a well-developed geometry involving no contradictions. To gain a partial and symbolic idea of its meaning, resort must be had to analogy with dimensions of a lower order.

An aggregate is said to be one, two, or three-dimensional according as one, two, or three numbers are necessary to determine any one of its elements. Considering space as an aggregate of points, a line is a one-dimensional space, because to determine the position of any point on it one number, giving its distance from some fixed point, suffices. Similarly, a plane is a two-dimensional space, and the point aggregate of ordinary space is three-dimensional. Thus, the exact position of any point of the earth is known when its latitude, longitude, and elevation above sea level are given. Now, if we have four variable, related quantities, each capable of assuming, independently of the others, every possible numerical value, we obtain a four-dimensional aggregate. Such an aggregate, if of points, constitutes four-dimensional space.

If we connect all points of our space (a 3-space)
with an assumed point outside of it, then the aggregate of all the points of the connecting lines constitutes a 4-space (hyperspace). Again, just as a point
moving generates a line, just as a line moving outside itself generates a surface, and a surface moving
outside itself generates a solid; so a solid moving outside of our space generates a hypersolid, or portion
of hyperspace. Or hyperspace itself may be conceived
as generated by our entire space moving parallel to
itself in a direction not contained in itself, just as our
space may be generated by the similar motion of an
unlimited plane, which may itself be generated by an
unlimited right line. Any space is that which forms
the boundary between two portions of a higher space,
and just as every unlimited plane divides our space
into two equal infinite parts, so every 3-space divides
hyperspace into two equal infinite regions between
which that 3-space forms a boundary of an infinitely
small thickness in the fourth dimension.

Two plane figures (say triangles) if in the same
plane may partially coalesce, but cannot intersect unless in different planes; similarly two volumes (say
cubes) if in the same 3-space may partially coalesce
but cannot intersect unless in different 3-spaces. In
hyperspace we have the following possible intersections: A hypersolid and a 3-space intersect in a solid,
two 3-spaces in a plane, three 3-spaces in a right line,
four 3-spaces in a point, a 3-space and a plane in a
right line, a 3-space and a right line in a point, and
two planes in a point. If the intersections are at an
infinite distance the intersecting elements are said to
be parallel, and if two 3-spaces are parallel all figures
or volumes in one 3-space are at equal distances from

the other 3-space. In the case of planes there are two kinds of parallelism, and parallel planes are either completely or incompletely parallel according as they are in the same or different 3-spaces, or as their intersection at infinity is a right line or a point.

To a given right line at a given point one can erect in a plane but one perpendicular, while in a 3-space one can erect an infinite number of perpendiculars, forming together a perpendicular plane, and in hyperspace an infinite number of perpendicular planes forming together a 3-space perpendicular to the given right line. A 3-space can also be perpendicular to a plane or to another 3-space. Planes may be perpendicular in two ways, incompletely or completely perpendicular, according as they are in the same 3-space or not; in the latter case every right line of either plane is perpendicular to every right line of the other.

The position of a point in a plane may be determined by its distance from each of 2 perpendicular right lines; in our space, by its distance from each of 3 mutually perpendicular planes; and in hyperspace, by its distance from each of 4 mutually perpendicular 3-spaces. In hyperspace these distances are accordingly measured along 4 mutually perpendicular right lines, which, taken by twos, determine 6 mutually perpendicular planes; and, taken by threes, determine the above-mentioned 4 mutually perpendicular 3-spaces. Just as in our space it requires at least 3 points to determine a plane, so in hyperspace it requires at least 4 points to determine a 3-space. A 3-space may also be determined by 2 non-intersecting right lines or by a plane and one point not in it.

Just as portions of our space are bounded by surfaces, plane or curved, so portions of hyperspace are bounded by hypersurfaces (three-dimensional), i. e., flat or curved 3-spaces. Hyperspace contains not only an infinite number of flat 3-spaces like ours but also an infinite number of curved 3-spaces or hypersurfaces of different types. A hypersphere, for instance, is a closed hypersurface all the points of which are equally distant from its center. Five points not in the same 3-space determine it, just as 4 points not in the same plane determine a sphere, and 3 points not in the same straight line a circle. All of its plane intersections are circles, all of its space intersections are spheres. A hypersphere of radius R passing through our space would appear as a sphere with a radius gradually increasing from zero to R and then gradually decreasing from R to zero.

While in our space there are but 5 regular polyhedrons (solids bounded by equal regular polygons), namely, the tetrahedron, cube, octahedron, dodecahedron, and icosahedron; in hyperspace there are 6 regular hyper-solids (cells), bounded by equal regular polyhedrons. These are C_5 (bounded by 5 tetrahedrons), C_8 (by 8 cubes), C_{16} (by 16 tetrahedrons), C_{24} (by 24 octahedrons), C_{120} (by 120 dodecahedrons), and C_{600} (by 600 tetrahedrons). All of them have been exhaustively studied by mathematicians, and models of their projections on our space have been constructed. Of these, C_8 (or the hyper-cube) is the simplest, because, though with more bounding solids than C_5, it is right-angled throughout, and therefore the standard form for measuring hyperspace. It is generated by a cube moving in the direction perpendicular to our space for a distance equal to one of its sides. In Fig. 1 where all dotted lines are supposed to be in hyperspace the initial cube is symbolically represented by $A\ B\ C\ D$

E F G H and the final cube by *A′ B′ C′ D′ E′ F′ G′ H′*, the direction *AA′* being supposed perpendicular to our space. Projecting* the edges of a hypercube on our space we get a network model of which Fig. 2 is a plane projection. The eight bounding cubes are represented in the model by the following projections: (1, 2, 3, 4, 5, 6, 7, 8), (5, 6, 7, 8, 9, 10, 11, 12), (9, 10, 11, 12, 13, 14, 15, 16), (13, 14, 15, 16, 1, 2, 3, 4), (1, 5, 9, 13, 2, 6, 10, 14), (2, 6, 10, 14, 3, 7, 11, 15), (3, 7, 11, 15, 4, 8, 12, 16), (4, 8, 12, 16, 5, 9, 13, 1). The form of the hypercube is conditioned by the mutual relations of these cubes that form its boundaries merely, as it contains an infinite number of cubes, just as a cube contains an infinite number of squares. In generating a hypercube by the motion of a cube, the latter's corners generate edges, its edges generate faces (squares) and its faces generate cubes. The resulting number of elements of the hypercube are therefore:

	In Initial Cube.	Generated.	In Final Cube.	In Hypercube.
Corners	8	..	8	16
Edges	12	8	12	32
Faces (squares).	6	12	6	24
Cubes	1	6	1	8

Each corner is common to 4 mutually perpendicular edges, to 6 faces and to 4 cubes; each edge is common to 3 faces and 3 cubes; and each face is common to 2 cubes. Every cube therefore has one face in common with 6 of the 7 others. We must conceive of the hypercube as composed of cubes starting from squares parallel to the faces of the cube and of these cubes all that exist in our space are the parallel squares from which they start.

---

* Not perpendicularly but as from a point at a little distance.—H. P. M.

In a plane the only kind of rotation possible is that about a point, in 3-space rotation can take place about an axis line, and in hyperspace about an axis plane.

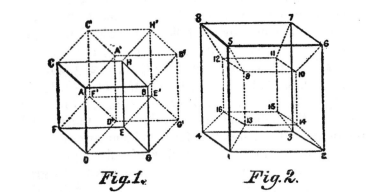

Fig. 1.   Fig. 2.

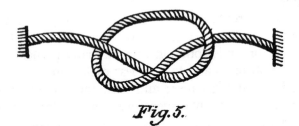

Fig. 3.   Fig. 4.

Fig. 5.

Two symmetrical plane figures such as the triangles *A* and *B* (Fig. 3) cannot be made to coincide by any movements in their plane, but by rotating one of them 180 deg. in the third dimension, it can be made to coincide with the other. Similarly, two symmetrical volumes (with faces equal but in reverse order) such as the hollow pyramids C and D (Fig. 4) cannot be made to coincide by any movements in our space, but by rotating one of them 180 deg. in hyperspace this can be done. The rotating pyramid disappears from our space, and upon its return after rotating 180 deg. it can be slipped into the other. In our space two movements of rotation will combine into a single resultant rotation, similar to its component rotations except that the direction of the axis is different. In hyperspace, however, there is in general no resultant for two rotations. Hence there are two different types of rotation in hyperspace and a body subject to two rotations is in a totally different condition from that which it is in when subject to one only. When subject to one rotation a whole plane of the body is stationary. When subject to the double rotation no part of the body is stationary except the point containing the two planes of rotation; and if the two rotations are equal, every point in the body, except that one, describes a circle.

Freedom of movement is greater in hyperspace than in our space. The degrees of freedom of a rigid body in our space are 6, namely, 3 translations along and 3 rotations about 3 axes, while the fixing of 3 of its points can prevent all movement. In hyperspace, however, with 3 of its points fixed it could still rotate about the plane passing through those points. A rigid body has 10 possible different movements in hyperspace, namely, 4 translations along 4 axes, and 6 rotations about 6 planes, while at least 4 of its points must be fixed to prevent all movement.

In hyperspace, a sphere if flexible could without

stretching or tearing be turned inside out. Two rings of a chain could be separated without breakage. Our knots would be useless. Thus the knot shown in Fig. 5 could be unknotted without removing the fastened ends. Just as in our space a point can pass in and out of a circle without touching its circumference, so in hyperspace a body could pass in and out of a sphere (or other inclosed space) without going through the surface surrounding it. In short, all of our space including the interior of the densest solids is open to inspection and manipulation from the fourth dimension, which extends in an unimaginable direction from every point of space.

Has hyperspace a real physical existence? If so, our universe must have a small thickness in the fourth dimension, otherwise like the geometrical plane assumed to be without thickness, our world too would be a mere abstraction (as indeed some idealistic philosophers have maintained), that is, nothing but "a shadow cast by a more real four-dimensional world." The real existence of a slight extension in the fourth dimension would, moreover, simplify certain scientific theories. For instance, in our space 4 is the greatest number of points whose mutual distances, 6 in number, are all independent of each other; but in hyperspace the 10 distances between any 2 of 5 points are geometrically independent. If this greater freedom of position were permissible to atoms, it would help to explain such chemical phenomena as isomerism, where molecules of identical composition have different properties. Again, rotation in hyperspace would explain the change of a body producing a right-handed into one producing a left-handed polarization of light. Further, Prof. McKendrick said before the British Association: "It is conceivable that life may be the transmission to dead matter . . . of a form of motion *sui generis*." Hyperspace has been brought somewhat into disrepute because spiritualists have assumed its existence in order to give "a local habitation" to their vagaries. Nevertheless, the possibility of its existence has not yet been shown to be inconsistent with any scientific fact, and the limitation of space to three dimensions, though probably correct, is therefore purely empirical.

Of what use then is the conception of hyperspace? For one thing, it gives a deeper insight into geometry. Thus, a circle considered merely as a one-dimensional aggregate of points has very few properties, while in a plane it has a center, radii, tangents, etc., and in 3-space has further numerous geometrical relations with the sphere, cone, etc. Similarly, the properties of any given line or surface increase in number when investigated in hyperspace. Also, just as it requires a 3-space to include certain one-dimensional aggregates (the helix, for instance), so in hyperspace hitherto unknown lines and surfaces become mathematically possible. Lower spaces are contained in higher (if curved, not necessarily the next higher) ; and just as the comprehension of plane geometry is enlarged by viewing plane figures in 3-space, so solid geometry is much illuminated by the geometry of hyperspace. Fields of mathematics hitherto inaccessible to geometry are now elucidated by geometrical representations. Finally, this conception effects a complete divorce between geometric space and real space, no longer considered necessarily identical, and in other ways also enlarges our mental horizon.

II.

NON-EUCLIDEAN GEOMETRY OF THE FOURTH DIMENSION.*

BY LIEUT.-COL. GRAHAM DENBY FITCH,

CORPS OF ENGINEERS, U. S. A.

The Fourth Dimension is an offshoot of the so-called "non-Euclidean" geometry, which has thrown so much light on the foundations of mathematics and on the nature of space.

For over 2,000 years Euclid was considered unassailable. His axioms were regarded as indisputable laws of real space, and his theorems as rigidly logical deductions therefrom. Neither view is correct. His axioms are metaphysical assumptions, and his theorems do not follow from them alone. The foundation of his method consists in establishing by superposition the congruence of lines, angles, plane figures, etc., and proof with him is thus merely a matter of constructive intuition. The axiom of "free mobility" (i. e., the possibility of moving figures in space without change of size or shape), which for instance does not hold on an egg-shaped surface but is essential to any system of geometrical measurement, is assumed by Euclid without being stated. (Hilbert discards proof by superposition, for motion itself needs a geometric foundation, and so cannot be a foundation for geometry.) Another of Euclid's tacit assumptions is that the straight line can be infinitely extended, which, true of Euclidean, is false of some non-Euclidean geometries (e. g. Riemann's).

---

* This supplementary essay was written by the winner of the prize after the award was made. It is here published as a historical *résumé* of the subject.

Euclid proves that "if alternate angles are equal, then the lines are parallel," but of the converse propositions,

"If alternate angles are unequal, the lines meet."

"If the lines are parallel, alternate angles are equal" (either of which implies the other) he could prove neither, and hence assumed the first, his celebrated fifth postulate, without which he could not proceed, as it was needed to prove the early important theorem that the sum of the angles of a triangle is not *less* than two right angles. This postulate of parallels appeared to later mathematicians neither self-evident nor independent of the other axioms. Considered a flaw, fruitless efforts were made for centuries to prove it. Yet here Euclid is right; this axiom or some equivalent (e. g. two intersecting lines cannot both be parallel to the same line) is necessary to Euclidean geometry.

It was from endeavors to improve upon Euclid's theory of parallels that non-Euclidean geometry arose. If the fifth postulate is really involved in Euclid's other assumptions, its denial must lead to contradictions; but about 1830 the Russian Lobachevsky and the Hungarian Bolyai, independently of each other, showed that its denial led to a system of two-dimensional geometry as self-consistent as Euclid's. This new geometry is based on the assumption that through a given point *a number* of straight lines can be drawn parallel to a given straight line.

Euclid's proof that the sum of the three angles of a triangle is not *greater* than two right angles was still considered perfect until the German mathematician Riemann in 1854 showed that it must involve a fallacy, because no premises were used not as true of spherical as of plane triangles, yet the conclusion is

false of spherical triangles. On this basis Riemann
further showed that still another self-consistent geometry of two dimensions can be constructed, based on the
assumption that through a given point *no* straight line
can be drawn parallel to a given straight line.

Thus we have three self-consistent geometries of
two dimensions, inconsistent as a rule, however, with
one another.

Let $P C$ (Fig. 1) rotate counter-clockwise about P.
Three different results are logically possible. When
the rotating line ceases to intersect the fixed line on the
right, either it will immediately intersect it on the left,

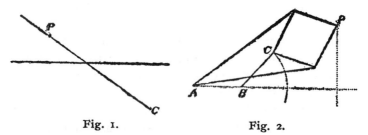

Fig. 1.   Fig. 2.

or it will continue to rotate for a time before intersection on the left occurs, or, lastly, it will intersect the
fixed line on both sides for a period of time. The first
possibility gives Euclid's, the second Lobachevsky's,
and the third Riemann's geometry.

The straight line of one geometry is not the same as
the straight line of another, but in all three geometries
it is the shortest distance between two points. Such
straightest lines are known as geodetic lines. Inciden-
tally it may be mentioned that the ordinary straight line
could not be drawn until recently except by means of a
straight edge. This of course was equivalent to assuming it. A method of constructing it was not discovered

until 1864, when a Frenchman, M. Peaucellier, first solved the problem, by means of a mechanism of seven links. This consists (Fig. 2) of two long links of equal length pivoted at the fixed point A, with their other ends pivoted to the opposite angles of a rhombus of four equal shorter links pivoted together, and of a final link pivoted to one free end C of the rhombus and to a fixed point B, the distance AB being equal to the link BC. If now C be made to describe a circular arc about B, P will describe a straight line perpendicular to AB, as can be readily proved by elementary geometry.

Defining space as "any unbounded continuum of geometric entities," the two non-Euclidean geometries, though logically on a par with the Euclidean, were considered inconsistent with reality until a space was known for which they held true. It was found, however, that Riemann's geometry is none other than that of a spherical surface (a two-dimensional space of constant positive curvature) provided arcs of great circles be taken as geodetic (straightest) lines. In 1868 the Italian Beltrami discovered a surface for which Lobachevsky's geometry held true, the so-called pseudospherical surface of infinite extent (a two-dimensional space of constant negative curvature). In our space only limited strips of the pseudo-sphere can be connectedly represented. It is a saddle-shaped surface (like the inner surface of a solid ring) ; and as the principal curvatures have their concavities turned in opposite ways, the curvature is negative. Euclid's geometry being true for the plane (a two-dimensional space of zero curvature), it will be seen that all three geometries require a space of constant curvature. On a pseudo-sphere the straightest line has two separate points at infinity, in a plane one, and on a sphere none.

Euclid's axiom that two straight lines—or more generally two geodetic lines—include no space, does not hold for geometry on the sphere. Euclid's fifth postulate, that two straight—i. e. geodetic—lines intersect when the sum of the interior angles is less than two right angles, does not hold for geometry on the pseudosphere. It can now be seen that Euclid's fifth postulate does not require nor admit of proof, because it embodies the definition of the kind of space to be dealt with (that of ordinary geometry).

Riemann also showed that there are logically three kinds of space of three dimensions, with properties analogous to the two-dimensional spaces mentioned. They are distinguished by a so-called measure of space curvature (purely analytical, not denoting curvature for sense perception). If this constant is zero, we have Euclidean space; if positive, we have spherical space; and if negative, we have pseudo-spherical space. In spherical space, straightest lines return upon themselves, and the back of our own head would be the ultimate background. Space would be unlimited but not infinitely great, and the sum of the angles of a triangle would exceed two right angles by an amount proportional to the area. In pseudo-spherical space straightest lines run to infinity as in Euclidean space, but the sum of the angles of a triangle is less than two right angles by an amount proportional to the area. In both spherical and pseudo-spherical space there are no similar figures of unequal size, for in each case the greater triangle must have different angles.

Lie proved that free motion can occur only in the above three spaces. There are other forms of non-Euclidean space which do not allow of free motion, called by Killing the Clifford-Klein spaces.

With three different self-consistent geometries of co-ordinate rank for the investigation of the properties of three-dimensional aggregates of points, it was natural to regard a space of any type as a locus in space of a higher dimension, and this led to the consideration of space of four dimensions, the properties of which, when of zero curvature, have been discussed in the main essay.

Euclidean space considered as a region of measureable quantities does not, as we have seen, correspond with the most general conception of an aggregate of three dimensions, but involves also special conditions. It is specially characterized by: 1, free mobility of rigid figures; 2, the single geodetic line between two points; 3, the existence of parallels; or by 1, free mobility; 2, postulate of similarity. Now these characteristics are not necessities of thought, and if they hold of real physical space, the fact must be determined by experience as in other empirical investigations, that is, by observation and experiment, for we cannot logically demand that the objective world must conform with our subjective intuitions.

It can never be proven, however, that our space is accurately Euclidean, for unavoidable errors of observation must always leave a slight margin in our measurements; and though within those limits our space is apparently Euclidean, this proves merely that the space constant is small, but not that it is zero. In spherical and pseudo-spherical triangles, the greater the area of the triangle the greater the difference of its angle sum from two right angles. But as even the interstellar triangles of parallax investigations are utterly insignificant compared with the dimensions of space itself, it must always remain an open question whether, if we had triangles large enough, the sum of the angles would still be two right angles. Even with our imperfect measurements, real space, however, might conceivably be proven to be Lobachevsky's (or Riemann's); for instance, if angular measurement could be made accurate to one millionth of a second, and if a lack (or excess) of two millionths were then found in the angle sum of some interstellar triangle.

Real physical space cannot be said to be either Euclidean or non-Euclidean. Geometry therefore throws no light on the nature of real space. The study of real space is an empirical science, while geometry is a construction of pure thought, a branch of pure mathematics. Pure mathematics is a collection of hypothetical, deductive theories, each consisting of a definite system of primitive, *undefined,* concepts or symbols and primitive, *unproved,* but self-consistent assumptions (commonly called axioms) together with their logically deducible consequences following by rigidly deductive processes without appeal to intuition. Pure mathematics thus reveals itself as nothing but symbolic or formal logic. It is concerned with implications, not applications. On the other hand, natural science, which is empirical and ultimately dependent upon observation and experiment, and therefore incapable of absolute exactness, cannot become strictly mathematical. The certainty of geometry is thus merely the certainty with which conclusions follow from non-contradictory premises. As to whether these conclusions are true of the material world or not, pure mathematics is indifferent. As applied, geometry, in short, is not certain, but useful.

The fact that all pure mathematics, including geometry, is rigidly deductive, is in fact nothing but formal logic, has important philosophical bearings. It definitely and finally refutes Kant, who based his entire philosophy on the supposed possibility of forming "synthetic judgments *a priori*"; that is, of obtaining absolute truth by the intuitions of pure reason quite independently of experience. For proof of his standpoint he referred to the existence of geometry. This argument was irrefutable until the discovery of non-Euclidean geometry. Another far-reaching conclusion is the following: Metaphysical axioms being mere imitations of geometrical axioms will, like the latter, have to be discarded. It seems therefore fitting to conclude with the following words of the eminent German mathematician Hilbert: "The most suggestive and notable achievement of the last century is the discovery of non-Euclidean geometry."

III.

FOURTH DIMENSION ABSURDITIES.*

BY "INCREDULUS ODI" (EDWARD H. CUTLER, A.M.,
NEWTON, MASS.)

The fourth dimension has no real existence in the
sense in which the external world that we know by

---

* First Honorably Mentioned Essay.—This author is attacking arguments
offered in proof of the existence of a space of four dimensions. Writers en-
thusiastic on the subject have given us details of the four-dimensional geometry and have tried to explain certain real or alleged phenomena by theories
based on this geometry. But the possibility of constructing a consistent system does not prove its existence and we may very well say in answer to these
writers that no experience has justified a belief in such existence and that no
well-authenticated facts are explained by these theories any more satisfactorily than by other theories. Some details, however, have been slightly mis-
understood by the author, or by the four-dimensional writers whom he is
answering, and his essay ought not to go out as an explanation of the fourth
dimension without a correction of his statements on these points.

He refers first to analogies drawn from the suppositions of a space of one
dimension and a space of two dimensions and our relation to the inhabitants
of such spaces. The analogies derived from line and plane geometries and
the relation of geometry of three dimensions to these geometries are very
useful in helping us form a conception of the four-dimensional geometry.
We may even apply these processes to physical conceptions and think of two-
dimensional and four-dimensional matter with a two-dimensional and a four-
dimensional physics. Thus two-dimensional matter in a two-dimensional
space might be impenetrable, one portion furnishing obstruction to the movements of another portion. While these suppositions may "furnish no basis
for belief in a fourth dimension," we should not say that they "involve a fatal
confusion of mathematical with physical conceptions." The question of the
existence or non-existence of such matter is a question of experimental physics
rather than a question of possible physical conceptions.

In speaking of lines, squares, and cubes and their boundaries, and of the
analogy by which we derive a conception of magnitudes in space of four dimensions bounded by solids, he says that there is no such analogy ; for "the only
possible boundary of a solid is a surface, whatever be the number of the dimensions of space." Apparently, he supposes that the magnitudes which are
bounded by solids are themselves solids, whereas they are portions of the
space of four dimensions. Lines are one-dimensional, surfaces are two-
dimensional, and solids are three-dimensional, whatever the number of

means of our senses has real existence. It is a philosophical and metaphysical conception, whose actual existence cannot be demonstrated by observation or by logical reasoning. The existence of the fourth dimension is regarded by some as in a high degree probable, and as furnishing a basis for metaphysical investigation, and a means of explaining some physical phenomena, the occurrence of which, however, is not universally admitted. It may also, like any supposition, true or false, be made the hypothesis for mathematical speculations, which are comprehensible, however, by the very small and select number only who are endowed

dimensions of space. This certainly gives us a "regular progression," leading, however, to something which is not a solid any more than a solid is a surface. We can think and reason about these figures although we may not be able to form any picture of them in our imagination.

Some writers have stated that a right glove turned into a left glove by rotation in space of four dimensions is turned inside out. This is not true and, of course, it cannot be explained, but the change from right to left produced by a simple rotation is easily explained, and indeed it is exactly analogous to the case of symmetrical triangles in a plane. This matter is discussed quite fully in the Introduction (p. 28).

If the space of our perceptions did lie within a space of four dimensions, then there would be a "new direction, not connected with any of those which we know, but at right angles to them all." Each direction which we know is at right angles to other directions which we know, but it does not follow that the new direction must coincide with them all or with any of them. We may "not need to be convinced that there is no such direction," but there is no confusion of thought in describing this direction.

Nor does the expression "entering the fourth dimension" seem to be "manifestly unintelligible," even if some slightly different phrase were better. If there were a "new direction" which we could not perceive, then our perceptions would not be unrestricted in direction. A body moving off in this direction would indeed "retain its length, breadth, and thickness," but would not remain within the range of our perceptions.

There is no question of the possibility in space of four dimensions of entering or passing out of what we call a tightly shut box or room, or of removing the contents of an egg without disturbing the shell (see foot-note p. 23). It is in this new direction that the walls of the room and the shell of the egg are supposed not to extend, and if such a direction did exist these movements would be possible without any modification of physical laws.

The space of our sensations and perceptions is only three-dimensional, but there is nowhere any contradiction in the Geometry of Four Dimensions, nor anything that is impossible.—H. P. M.

by nature with the ability to cope with original investigation in the domain of the higher mathematics.

The word "dimension" is more readily explained than defined. All more or less clearly conceive of space as extending indefinitely or infinitely in every direction; and of extension in space there are three "dimensions" —length, breadth, and thickness. Or, in another point of view, having three fixed points from which to reckon measurement, by three dimensions or measurements we can fix exactly the position of any point in space. Thus, if the three fixed points be the center of the earth, one of the poles, and some other point on the surface, as the location of the Royal Astronomical Observatory at Greenwich, the length of the line drawn from the center of the earth to the point in question in space, as a star, however remote, and the latitude and longitude of the point in which the line from the center intersects the surface, will be three dimensions, which fix exactly the position of the point in space, or of the star. Or again, starting from any point in space, we may reach any other point by proceeding successively in three directions at right angles with one another. Thus, moving from the starting point, first the proper distance east or west, then from the point arrived at the proper distance north or south, and finally the proper distance up or down, we reach the second point in question.

In all the ways in which the meaning of the word is thus illustrated we see that we can have no fewer and no more than three dimensions; but the believers in a fourth dimension infer its existence from analogy in one of the following deductive processes:

(1) Conceive, we are bidden, of a space of but one dimension. A being in such a space would be limited to a straight line, which he would conceive as extending

infinitely in both directions. His only possible movement would be along this line, and if he encountered another being, neither could pass the other. If he is really within a space like ours, although his perception is confined to one direction only, and a being in our space should lift one of the two beings, and place him on the other side of the first, the latter would lose sight of the other as soon as the lifting took place, and the movement by which the change of position had been effected would be utterly unintelligible to him.

Conceive of a space of but two dimensions, like the flat surface of a table. Beings in such a space could move around one another, but one of them completely surrounded by others would be imprisoned by them. If, as before, the two-dimension space is within our space, and really depends on the limitation of the perceptive faculties of the beings in question, the imprisoned being could be lifted by a being in our space, and set down outside of the beings surrounding him. The latter would lose sight of him during this movement, and not understand how it had been effected.

From these suppositions of one-dimension space and two-dimension space, the inference is drawn that there may be a fourth dimension in our space, and that our ignorance of it arises only from the limitation of our perceptive faculties.

These suppositions, however, involve a fatal confusion of mathematical with physical conceptions. Mathematical lines and plane figures do not, like matter, occupy space, and they present no obstruction to the movements of one another. They may freely intersect, or pass through one another, or coincide wholly or in part with one another. If these supposititious beings in one- or two-dimensional space find any obstruction to their movements, it must be because they occupy space, and therefore are really in three-dimension space, however little they extend except in one or two directions. A line or a plane surface can be conceived only with space around it in every direction. The supposition of a one-dimension or a two-dimension space is therefore impossible except as a mathematical abstraction, and furnishes no basis for belief in a fourth dimension.

(2) The straight line, a one-dimension magnitude, ends in points; the square, a two-dimension magnitude, is bounded by straight lines, one-dimension magnitudes; the cube, a three-dimension magnitude, is bounded by squares, two-dimension magnitudes. It is inferred by analogy that three-dimension magnitudes bound four-dimension magnitudes, although the latter are not known to us. Thus the "four-dimensional cube" receives a name, the "tesseract," and is said to be bounded by cubes.

But there is no such analogy as is here assumed. All lines end in points, although some lines, like circular arcs, require two-dimensional space, and others, like a corkscrew curve, three-dimensional. Nor are all two-dimensional figures bounded by straight lines. The bounding lines of circles and ellipses, for example, require two-dimensional space, as much as the figures themselves. Still further, solids like spheres or egg-shaped bodies, are bounded by three-dimension surfaces. There is, therefore, no regular progression which would lead us to suppose the existence of magnitudes bounded by solids. In fact such a supposition is inconceivable. The only possible boundary of a solid is a surface, whatever be the number of the dimensions of space.

(3) In the series of the successive powers of a

number, a, a^2, a^3, a^4 . . . a^n, a may be represented graphically by a straight line, of which a denotes the length; a^2, by a square, of which a denotes the length of a side; a^3, by a cube, of which a denotes the length of an edge. It is inferred that if we keep on, there must be a magnitude corresponding with a^4, and so on indefinitely up to a^n. Such magnitudes are incompatible with three-dimension space, and suggest for their possible existence "spaces of higher order."

To those who have some elementary knowledge of analytical geometry, or even of the use of graphs in algebra, the origin of the conception of spaces of higher order may be presented in a different way. As an equation containing two "variables" may be considered as representing the locus of a series of points in a plane, so an equation with three variables is the locus of points in space, referred to three rectangular axes. But since, as shown above, in explaining the word "dimension," three dimensions or co-ordinates fix definitely and exactly the position of a point, equations with more than three variables transcend the scope of our geometry, and require for analogous interpretation spaces of more than three dimensions.

There is no objection to the hypothesis of spaces of a "higher order" as a purely mathematical conception; but this abstract supposition has no bearing on the number of dimensions of actual space as we know it.

(4) If we connect by a straight line the vertex of an isosceles triangle with the middle point of the base, we have divided the triangle into two triangles which are plainly equal. If we were confined to the two-dimensional surface of which the triangles are a portion, we could never move them about so as to apply one to the other, and prove them equal by coincidence. Not

being under this restriction, but being in three-dimension space, we turn one of the triangles a half revolution on one of its sides, and then the two figures may be made to coincide. Now there are many symmetrical solids, for instance, the two hands, which can never be brought into identical shape. We cannot prove the left hand equal to the right by putting on the left the right-hand glove. But if we turn the right-hand glove inside out it will fit the left hand. Just as we can prove two-dimensional figures equal by availing ourselves of the possibilities presented by three-dimensional space, it is inferred that in four-dimensional space, not only the glove, but the hand within it, might be turned inside out, and made identical in shape with the other hand. No explanation is offered of the way in which an additional dimension would render such an eversion possible, and if we could admit that it would do this, we are not shown why the actual existence of a fourth dimension follows. Some four-dimension enthusiasts appear to believe that symmetrical forms in organic bodies could not originate without a fourth dimension, but no reason is given for this belief.

The four numbered sections above include virtually all the lines of thought along which the effort is made to substantiate the existence of a fourth dimension. Metaphysical considerations are sometimes added of the uncertainty and possible inaccuracy of our conception of space, but with no suggestion for correcting this inaccuracy, and no argument for the belief in a fourth dimension. Admit that the mind must itself contribute an *a priori* element to all knowledge, and that the truth of things is not limited by the phenomenal apprehension of them; it does not follow that this apprehension is to be assumed without demonstration to be false or incomplete. In an investigation like the present one it is unnecessary to consider whether our conception of the *non ego* is subjective or objective; we must accept the world of matter and of mind in which we live as our perceptions present it to us, and as it is generally conceived. No observation has ever discovered the existence of a fourth dimension in space, and it may safely be said that there is no reason for believing in its existence.

The theory of spaces of a higher order, as developed in section (3) above, is entirely legitimate as an abstract mathematical conception, but furnishes no basis for the supposition of a fourth dimension in our space. It virtually assumes space as we know it to be three-dimensional; yet from a suggestion arising from this theory apparently (for no other origin for the assumption is to be found) the four-dimensionists have made space as we know it a space of the highest order; for the same analogies and inferences on which they depend would lead us to a fifth, a sixth, an *n*th dimension. A fourth dimension belongs (or rather four dimensions belong) to the theoretical four-dimension space; but mathematics furnishes no basis for ascribing to our space more or fewer than three dimensions.

The confusion of thought of the four-dimensionists characterizes their writings on the subject. The most thorough-going devotee of the fourth dimension asserts: "There is nothing mysterious at all about it. . . . From every particle of matter there is a new direction, not connected with any of those which we know, but independent of all the paths we can draw in space, and at right angles to them all." It would seem indisputable that a direction at right angles with all the paths or lines that we can draw in space from any point, would produce lines coinciding with all the lines drawn from the point, and therefoie giving no "new direction." But we do not need to be convinced that there is no "direction" from which we are cut off, and in which we cannot direct our perceptions.

The attempted analogies described in section (1) above, are those on which the four-dimensionists chiefly depend, and they rely upon them to show that a fourth dimension would explain how a body may become invisible. They assert that a body would disappear on "entering the fourth dimension." This expression is manifestly unintelligible. Every body extends constantly in all the dimensions of space; we cannot think of it as "entering the dimension" of length, breadth, or thickness, or of "entering the fourth dimension," if there were one. But the disappearances produced as in section (1) depend wholly on removal from the limited perceptive faculties of the supposed observers; but our normal perceptions are unrestricted in direction, and extend to every point in space, unless cut off by distance or by an interposed physical obstruction. If all the particles of a body moved in the "new direction" of the imaginary fourth dimension, the body would still retain its length, breadth, and thickness, and would still remain within the range of our perceptions.

The assertion is made on the authority of eminent mathematicians, that in space of four dimensions there would be no obstruction to entering or emerging from space shut in on every side, as a tightly shut box or room, and "the fourth dimension" is relied upon to explain supposed mysterious occurrences of such entrance or emergence. The modification of physical laws in spaces of a higher order, those of unusual mathematical ability alone can be expected to understand, and in the special instance in question no explanation is vouchsafed. Until such explanation is given, those who can make no claims to exceptional mathematical talent will be unable to believe it possible, in space of the fourth, or of any order, to extract the contents of an egg, or to pass an object within the egg, and at the same time leave intact the continuous material structure that we call the shell. But whatever may be possible in theoretical spaces of higher order, we need not accept an unintelligible fourth dimension to aid in the explanation of something equally unintelligible.

It may be said in conclusion, that the only "explanation of the fourth dimension" that can reasonably be given, is to say that, in the sense in which the expression is used, the fourth dimension is absolutely non-existent. It could have meaning only to designate the dimension, in addition to the three that we know, belonging to the imaginary mathematical hypothesis of four-dimension space. The "fourth dimension" has no relation to the actual universe in which our sensations and perceptions are exercised, and belongs to that realm of thought to be entered only by the select few, whose exceptional genius includes the development of the mathematical imagination.

IV.

THE BOUNDARY OF THE FOUR-DIMENSIO-
NAL UNIT AND OTHER FEATURES OF
FOUR-DIMENSIONAL SPACE.*

BY "PLATONIDES."

The schoolboy early becomes familiar with linear measure, square measure, and solid or cubic measure. He understands them respectively as "the measurement of lengths," "the measurement of surface which depends on length and breadth taken conjointly," and "the measurement of volume which depends on length, breadth, and height all taken together." The first involves one dimension, length; the second, two mutually

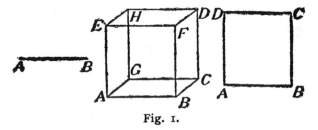

Fig. 1.

perpendicular dimensions, length and breadth, multiplied together; and the third, three dimensions, each perpendicular to the other two—length, breadth and height, all multiplied together. Let the units of these three kinds of measure (e. g., foot, square foot, and cubic foot) be represented by a line *AB,* a square *AB CD* with that line as side, and a cube *ABCD-G* with that line as edge and that square as base (Fig. 1).

* Second honorably mentioned essay.

The unit AB may be regarded as made up of an indefinitely large number M of points arranged continuously from A to B; the square $ABCD$ then contains $M \times M$ $= M^2$ points; and the cube $ABCD$-G contains $M \times M$ $\times M = M^3$ points. One can go from any point in AB to any or every other in AB by moving in the one fixed direction of AB; similarly, from any point to any or every other in $ABCD$ by moving in the two fixed directions of the bounding lines; and likewise in $ABCD$-G by moving in the three fixed directions of the bounding lines (direction forward or backward being regarded as the same in every case). Hence, with regard to motion from one point to another, the first unit is one-dimensional, the second, two-dimensional, and the third, three-dimensional.

Man can make no motion which cannot be resolved into a combination of three mutually perpendicular directions; he can reach no place which cannot be reached by going north or south, east or west, and upward or downward; he can find no point in a room which cannot be found by moving in the direction of the length, breadth, and height of the room. Sight reveals two dimensions directly, the breadth and the height of the object beheld, while the third dimension, the distance of the object, is estimated by means of the muscular turning of the eyes to focus them on it. No sense calls for a fourth direction, perpendicular to the other three; in fact, all of man's experience leaves him satisfied with three dimensions.

Leaving experience behind and reasoning wholly from analogy, the fourth dimension is introduced as follows: Four-dimensional measure depends on length, breadth, height, and a fourth dimension, all multiplied together. It involves four linear dimensions, each perpendicular to the other three; consequently the fourth dimension is at right angles to each of the three dimensions of the three-dimensional measure. Its unit must have AB as edge, the square $ABCD$ as face, and the cube $ABCD$-G as base. It contains $M \times M \times M \times M = M^4$ points. To travel from any point to any or every other point in it is possible by moving in the four fixed directions of its bounding lines.

The square $ABCD$ (Fig. 1) is derived from the line AB by letting AB with its M points move through a distance of one foot in a direction perpendicular to the one dimension of AB; every point of AB in this motion describes a line, and $ABCD$ contains, therefore, M lines, as well as M^2 points. The cube $ABCD$-G is derived from the square $ABCD$ by letting $ABCD$ move one foot in a direction perpendicular to its two dimensions; its M lines and M^2 points describe respectively M squares and M^2 lines; accordingly $ABCD$-G contains M squares, M^2 lines, and M^3 points. Similarly, the four-dimensional unit is derived from the cube $ABCD$-G by letting that cube move one foot in a direction perpendicular to each of its three dimensions, i. e., in the direction of the fourth dimension; its M squares, M^2 lines, and M^3 points describe respectively M cubes, M^2 squares, and M^3 lines; accordingly the four-dimensional unit contains M cubes, M^2 squares, M^3 lines, and M^4 points. Considering the boundaries of the units, AB has two bounding points, $ABCD$ has four, $ABCD$-G has eight—four each from the initial and the final positions of the moving square—and the four-dimensional unit has 16—eight each from the initial and the final positions of the moving cube. Of bounding lines, AB has one (or is itself one), $ABCD$ has four, $ABCD$-G has twelve—

four each from the initial and the final positions of the moving square, and four described by the four bounding points of that square; and the four-dimensional unit has 32—twelve each from the initial and the final positions of the moving cube, and eight described by the eight bounding points of that cube. Similarly, of bounding squares, *ABCD* has one (or is itself one), *ABCD-G* has six—one each from the initial and the final positions of *ABCD,* and four described by the bounding lines of the moving square—and the four-dimensional unit has 24—six each from the initial and the final positions of the moving cube and twelve described by the bounding lines of the moving cube. Finally, of bounding cubes, *ABCD-G* has one (or is itself one), and the four-dimensional unit has eight— one each from the initial and the final positions of the moving cube, and six described by the bounding squares of the moving cube.

If the bounding lines of the square *ABCD* are supposed to be made of a continuous wire and that wire is cut at *D,* the boundary may then be folded down into line with *AB,* forming a one-dimensional figure (Fig. 2) of four linear units. The original linear

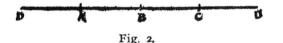

Fig. 2.

unit *AB* has one linear unit at either side of it and an extra one, *CD* beyond on one side. If the cube *ABCD-G* has its bounding squares supposedly made of a continuous sheet of tin and that sheet is cut along the lines *EF, GH, HE, AE, BF, CG,* and *DH,* the square faces can be folded down to form a two-dimensional figure of six squares. The square *ABCD* has a square at each side of it and an extra one, *EFGH,* beyond on one side (Fig. 3). Likewise, if the four-

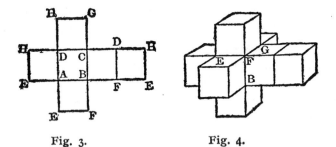

Fig. 3.　　　　　Fig. 4.

dimensional unit has its bounding cubes made of connected solid wood and this wood is cut through the appropriate planes, the cubes can be folded down to form, by analogy, a three-dimensional figure of eight cubes. The cube *ABCD-G* has a cube at each side of it and an extra one beyond on one side (Fig. 4). These eight cubes, now forming a three-dimensional figure, constituted the boundary of the four-dimensional unit.

The following table shows the results obtained for the contents and the boundaries of the four units considered:

CONTENTS.

	Points.	Lines.	Squares.	Cubes.
One-dimensional unit.....	M	1	0	0
Two-dimensional unit....	M^2	M	1	0
Three-dimensional unit...	M^3	M^2	M	1
Four-dimensional unit....	M^4	M^3	M^2	M

BOUNDARIES.

	Points.	Lines.	Squares.	Cubes.
One-dimensional unit....	2	1	0	0
Two-dimensional unit....	4	4	1	0
Three-dimensional unit.	8	12	6	1
Four-dimensional unit...	16	32	24	8

The reasoning used is capable of extension at once to units of five, or even more, dimensions.

If the one-dimensional unit is extended indefinitely to the right beyond B and to the left beyond A so that its length becomes greater than any number one can name, it represents a one-dimensional space. Similarly, the indefinitely great extension, equally in every dimension, of the other units gives a representation respectively of two-dimensional, three-dimensional, and four-dimensional spaces.

The one-dimensional unit is separated from the rest of the one-dimensional space in which it lies by two points, the two-dimensional unit from the rest of its two-dimensional space by four lines, the three-dimensional unit from the rest of its space by six squares, and, similarly, the four-dimensional unit is separated from the rest of the four-dimensional space in which it lies by eight cubes. To inclose an object of any number of dimensions in space of the same number of dimensions demands, in one-dimensional space, two points, in two-dimensional space, at least three lines, in three-dimensional space, at least four planes, and, in four-dimensional space, at least five three-dimensional spaces.

As with the units, so with the spaces, any point can be reached from any other in the same space by moving in as many fixed directions, each perpendicular to the rest, as that space has dimensions.

Time represents a one-dimensional space; for it proceeds in one direction only from an indefinitely remote past to an indefinitely distant future (Fig. 5).

Fig. 5.

The present is a point traveling through time (or allowing time to slip past it) with uniform velocity; and any point in time can be reached by traveling through a definite distance (in years, months, etc.) from one chosen fixed point (e. g., the birth of Christ).

Any portion of the earth's surface, regarded as a plane, represents a portion of a two-dimensional space; and the two fixed directions are those of latitude and longitude. An illustration of three-dimensional space is that space—to man's perception—in which the universe is placed. Man can find no illustration of a four-dimensional space.

If two lines, AB and $B'A'$, in the same one-dimensional space are symmetrical about a point O of that space (Fig. 6), AB cannot be so moved in that space

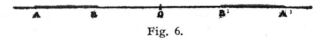

Fig. 6.

that the corresponding points shall coincide (A with A', B with B', etc.). To effect such coincidence, it is necessary to rotate AB through two-dimensional space about O as a center; or, roughly speaking, AB

must be taken up into two-dimensional space, turned over, and put down on $B'A'$. Likewise, if two triangles, in the same two-dimensional space, are symmetrical with respect to a line (Fig. 7), such coinci-

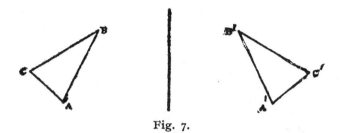

Fig. 7.

dence of corresponding points and lines can be effected only by rotating one triangle through three-dimensional space about the line of symmetry; or, roughly speaking, one triangle must be taken up into three-dimensional space, turned over, and put down on the

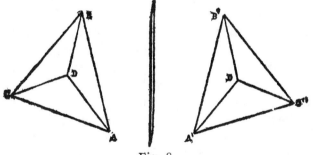

Fig. 8.

other. Again, if two polyhedral figures in the same three-dimensional space are symmetrical with respect to a plane (Fig. 8), coincidence of corresponding points, lines, and planes can be effected only by rotating one polyhedral figure through four-dimensional space about that plane; or, roughly speaking, one of the polyhedral figures must be taken up into four-dimensional space, turned over, and put down on the other. A right hand and its reflection (a left hand) in a mirror are symmetrical with respect to the plane of the mirror and rotation about that plane would effect coincidence. Such rotation would make a right glove become a left glove; or, roughly speaking, a right glove tossed up in the direction of the fourth dimension and turning over there will fall back a left glove.

The inability of man to locate the fourth dimension or to detect the existence of a four-dimensional space, even if it be close at hand, is comparable with the inability of a two-dimensional man, inhabiting a two-dimensional space, to locate the third dimension or to detect the existence of three-dimensional space, even though his own space might be part of it, as a plane is part of a solid. Suppose the two-dimensional space represented by this page to be inhabited by two-dimensional beings. They have length and breadth, can move in those two dimensions, and are supposedly conscious of them. They have no thickness, cannot rise from the paper or sink beneath it, and are unconscious of any dimension in such a direction; they have no "upward" and no "downward." Let them have intelligence concerning all within their space to the extent that man is intelligent regarding his universe; let them possess houses and barns, and in general let their life be as rich as may be. Their houses and barns will have no roofs and no floors, for the bounds of the space itself alone are there. Three lines are sufficient to inclose any object in their world, and the flat-man himself is exposed only along his polygonal contour; the interior of his polygon—his own interior—is to be reached only through his contour, for there is no "above" and no "below" within his cognizance. To convince him that a third dimension of "upward" and "downward" exists, touching and leading from even the interior of his polygon—his own internal parts—would be a hopeless task. Even if he accepts the arguments from analogy as to the properties of such a dimension, he would rebel at the idea of looking within himself to find it. Yet, even there, at right angles to the two dimensions which he knows, it is to be found —as well as everywhere else in his space. And, similarly, within himself, quite as much as anywhere else, must man look if he is to find the fourth dimension.

Were one to explain to this flat-man that a three-dimensional being, approaching from the direction of that unknown third dimension, could reach within his most securely locked barn and remove its contents without opening a door or breaking a wall—or could touch the very heart of the flat-man himself without piercing his skin—the flat-man might still be none the nearer to an appreciation of the third dimension. Equally impossible is it for man to understand from what direction a four-dimensional robber must come to steal the treasures from the soundest vault without opening or breaking it—or by what way of approach the four-dimensional physician would reach to touch the inmost spot of the human heart without piercing the skin of the body or the wall of the heart; yet the route of such a robber and of such a physician lies along the fourth dimension. By that route must come the four-dimensional being who is to remove the contents of the egg without puncturing the shell or drink the liquor from the bottle without drawing the cork. Such four-dimensional creatures, inhabiting a space containing the three-dimensional space where man lives, would constitute the most perfect of ghosts for man's world; and the absence of such ghosts argues against the existence of a four-dimensional space so situated and so inhabited.

Algebra demands that geometry picture all its problems; and since an algebraic problem may contain four or five or more unknown quantities quite as well as any lesser number, algebra demands a four-dimensional, five-dimensional, or higher space for its use quite as imperatively as the spaces of lower dimensions. Perhaps certain phenomena of molecular physics or the mechanical principles of the electric current may find a complete explanation only with the use of the fourth dimension. Perhaps the fourth dimension escapes man's discovery only because the measurements in its direction are always very minute in comparison with the measurements in the three other dimensions. Thus far, however, the space of four dimensions—and all spaces of more dimensions —may be only "the fictitious geometric representation of an algebraic identity."

V.

HOW THE FOURTH DIMENSION MAY BE STUDIED.*

BY "CHARLES HENRY SMITH" (CARL A. RICHMOND, CHICAGO, ILL.).

A colony of bees housed in a hive with glass walls so that their every movement can be observed affords a very instructive lesson in natural history. Such a glass hive may also serve as a helpful illustration in a consideration of the fourth dimension. Let us imagine a hive with its floor and roof of horizontal glass plates brought so close together that there is barely room for the bees to move about between them, and, for the purpose of our illustration, let us endow the bees with the intelligence of men. To these bees, so confined, forward and backward, right and left, would be familiar ideas and their world would be one of two dimensions only. Debarred from upward and downward movement by the closeness of the glass plates, the words "up" and "down" would be meaningless to them because there would be no experience upon which to base these ideas. Imperfect as is the illustration, it suggests the conception of a world of only two dimensions, length and breadth.

Plane geometry is a science which deals with such figures as triangles, squares, and circles. It is interesting to know that it originated in Egypt where it was developed to facilitate the measurement of land.

---

*Third Honorably Mentioned Essay.

This origin of the science gave rise to the name geometry, which means earth measurement. Long subsequent to the era of its Egyptian development the science was extended under the names of solid geometry to a study of such figures as spheres, cubes, and cones.

The bees in the glass hive could move around a square, could make triangles and circles, and to them plane geometry would be a practical science; but with their ignorance of an up-and-down direction, a cube or sphere would be inconceivable, and a third dimension would appear to them as absurd and unthinkable as a fourth dimension does to us. Suppose we lay two pencils on the table so as to cross one another at a right angle and then hold a third pencil so as to form right angles with the other two. While this is obviously a possible thing for us to do, it would be impossible for the bees with their ignorance of the dimension of height. They could, of course, have two slender pencils in their hive at a right angle to one another, but they could not have a third pencil at right angles to both of the first two. We may look upon the two pencils as representing the two dimensions of the world of the bees, and the three pencils as representing the three dimensions of our world. Suppose, further, that some one tells us to hold a fourth pencil at right angles with the other three. In our field of experience we can find no place for it, just as the bees could find no place in their field of experience for the third pencil. This fourth pencil represents the so-called fourth dimension. Although it is impossible for us to place it, the illustration of the relation of the bees to the third pencil or dimension teaches us that the limitations of experience ought not to be deemed conclusive as to how many dimensions space may have.

It is a matter of pure speculation as to whether there is such a thing as a fourth dimension, whether there are beings of intelligence to whom phenomena are manifested in the form of four dimensions. It is by no means the attitude of mathematicians instantly to recoil from the suggestion, but they are pleased to go ahead and study as accurately as possible under the necessary limitations what may be the properties of a space of four dimensions, if there is any such thing. The fundamental guiding principle of their investigation is this: Whatever they find to be the relations of geometry of two dimensions to geometry of three dimensions, they assume that there are similar or analogous relations between geometry of three dimensions and geometry of four dimensions. As the circle is to the sphere, so is the sphere to some unknown body, which may have its existence in space of four dimensions. As the square is to the cube so is the cube to a figure in space of four dimensions which we may call the "cuboid."

Of course the fourth dimension is intangible. Mathematicians do not ask us to imagine a fourth dimension, much less do they ask us to believe in it. It is not to be supposed that the most skilled student in this subject has a mental picture of four-dimensional space. Nevertheless, the properties and relations of figures existing in four-dimensional space may be investigated and stated.

Algebra is the science of numbers. It is a very efficient aid in the study of geometry. Algebra deals largely with equations such as $x y = 12$, which means that x and y are two variable numbers that multiplied together, give 12, as for example, 3 and 4 or 5 and 2 2/5. All the simpler figures of geometry such as the straight line and the circle may be represented by equations; in other words, the equations are condensed descriptions of the respective geometrical figures, somewhat as a score-card is a condensed description of a base-ball game. Mathematicians have learned that the properties of geometrical figures can be studied far more readily by means of their equations than by means of the figures themselves. A mathematician who understands this mode of study can look at the equation of a curve and tell all sorts of interesting and useful properties of it without ever seeing the curve itself—indeed, without even having any mental picture of what the form of the curve may be.

Without going into detail, it may be stated that one equation with two variable numbers represents a plane figure, thus $x^2 + y^2 = 15$ represents a circle. One equation with three variable numbers represents a figure in space, thus $x^2 + y^2 - z^2 = 0$ represents a cone. What does one equation with four variable numbers represent, say, for example, $x^2 + y^2 + z^2 + w^2 = 20$? By analogy, we should say a figure in space of four dimensions. Althought we cannot imagine such a thing, we can pursue our analogies and study this unreal figure by means of its equation, and thus we can deduce many of its properties. The difference is simply this: whereas, when we study the equation of a cone, we can always turn to the real cone and interpret our results thereon, when we study an equation of a four-dimensional figure we have to be satisfied without such an interpretation. In other words, although our geometry halts with three dimensions our algebra marches on to any number of dimensions and is a stimulus to imagine a geometry of more than three dimensions.

We will now outline briefly a way in which algebra may help to give a person some faint notion of a figure having four dimensions. It is somewhat common to study a figure having three dimensions by means of equally spaced parallel sections thereof. For example, if the microscopist wants to study the shape and structure of a germ cell, he slices off exceedingly thin sections and arranges them in succession on a glass slide. Then by looking at these sections in succession he can form an idea of the solid structure of the germ cell. Mathematicians have rules by which such sections of a solid figure may be constructed by means of equations. They start with an equation which represents a solid body, for example, $x^2 + y^2 + z^2 = 9$ representing a sphere, and they perform certain operations by which they get a series of resulting equations that represent the successive sections of the solid body. It remains, then, merely to draw pictures of the sections from the data afforded by the resulting equations. By looking at all these pictures, a person may get a fair idea of the shape of the original solid. In the case of a sphere the sections are circles of varying size. As we have already stated, an equation having four variable numbers, should by analogy represent a figure in space of four dimensions. Suppose we have such an equation, as $x^2 + y^2 + z^2 + w^2 = 20$. We can apply the same rules and perform the same operations to get sections of the figure represented by this equation. Curiously but consistently, these sections come out as solid figures. From the data afforded by the resulting equations, the mathematician can model these solid figures in clay and lay them in a row on the table before him. Just as the microscopist looks at the series of sections on his slide to get an idea of the solid structure of the germ cell, so the mathematician can look at the series of clay models before him and possibly feel that he has some idea of the nature of the four-dimensional figure represented by the equation with which he started.

Thus we see how the fourth dimension may be studied by means of the equations which algebra furnishes. There is another bolder way. We have seen that we can hold three pencils so that each one of them will make a right angle with each of the others. Instead of saying that it is absurd to suppose that a fourth pencil can be held in a position so as to form right angles with each of the first three pencils, let us *assume* that it *can* be done. Without any further assumptions a complete geometry of four dimensions can be built up by pure reasoning. Many of its conclusions are no more obvious to the senses than is the fundamental assumption with which it starts. Still that is the *only* assumption; all else may be deduced from that one assumption and from the principles of our well-known plane and solid geometry.

An illustration of a special method in the study of space of four dimensions may serve to show how mathematicians reason about such things without being able actually to imagine them. We proceed by ascertaining the relations between two dimensions and three dimensions, and then establishing these relations by analogy between three dimensions and four dimensions. Suppose we have a glass cube resting on the table before us and we close one eye and look straight down upon it with the open eye. Its appearance will be as shown in the accompanying drawing. This drawing is really a plane figure, of two dimensions, and it might have been produced in the following manner; namely, by drawing one square inside of another and then drawing lines connecting the corresponding corners. All this could be done without any

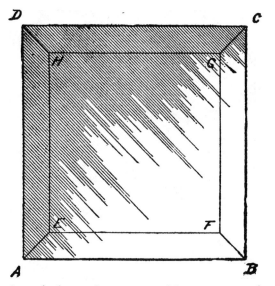

Top view of glass cube as seen with one eye : a three-dimensional figure appearing in one plane.

thought of three dimensions. The bees in the glass hive could draw such a figure as the one here on the paper before us. Nevertheless, on this figure many of the properties of the cube can be studied. By counting the four-sided figures (*ABCD, EFGH, AEFB, BFGC, CGHD, DHEA*), which we find to be six, we learn how many faces the cube has. By counting the corner points, which are eight, we learn how many

corners the cube has.   By counting the lines, which
are twelve, we learn how many edges the cube has.
Just as starting with the squares we are able to get a
two-dimensional figure, which, for the purpose of
investigation, may be taken as representing the cube,
may it not be possible that starting with cubes we

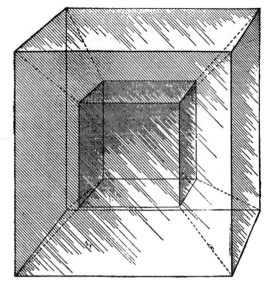

Analogous view of a "cuboid" of four dimensions appearing
as a figure of three dimensions.

can get a three-dimensional figure which shall represent the four-dimensional figure which we call the
cuboid?   Just as we drew a smaller square inside of
a larger one, so we should think of a smaller cube
inside of a larger cube, and just as we drew lines
joining the corresponding corners in the case of the
squares, so we should make planes joining corresponding edges in the case of the cubes.   The figure

so formed is somewhat imperfectly pictured in the accompanying drawing, and for the sake of clearness, let us suppose we have such a solid glass figure before us. In the case of the squares, to find from them how many square faces the cube has, we counted the big outer square, the small inner square and the four surrounding figures and got six as the result. So in the case of the cubes, to find from them how many cube faces the cuboid has, we count the big outer cube, the small inner cube and the six surrounding solid bodies and thus get eight as the result; this indicates that the cuboid has eight cube faces. A further study of this representative figure discovers that the cuboid has 24 plane square faces, 32 edges, and 16 corner points. This shows how we can get a representation of a four-dimensional body, and on this representation we can study its properties. There are many considerations which we have not space to present which confirm the accuracy of the deductions that have just been stated.

What is the use of such generalities, abstractions and speculations? About the same as to know whether the earth goes around the sun or the sun goes around the earth. Space is as properly an object of scientific study as are planets or geological strata. Moreover, the study of these fundamental things in geometry throws light on the nature of our own mental equipment. We learn better what is the nature of reasoning processes and how knowledge is built up from simpler and more fundamental elements. Such speculations sometimes lead to very useful results.

If you hold 5 marbles in your hand and are told to take away 8 of them, this suggestion seems as unthinkable as the suggestion of a fourth dimension. But when men chose to represent by —3 the result of subtracting 8 from 5, instead of simply saying it was impossible, then the foundation was laid for the enormously useful science of Algebra.

The assumption of a fourth dimension has not as yet led to any noteworthy useful results, but it is by no means impossible that the science of four-dimensional geometry may come to have useful applications. It has been suggested by Prof. Karl Pearson that an atom may be a place where ether is flowing into our space from a space of four dimensions. It can be shown mathematically that this would explain many of the phenomena of matter. At the present stage, the suggestion is regarded, even by its author, as merely fanciful, though it is not as fanciful as the proposition of the German spiritualists who regard the fourth dimension as the abode of their disembodied spirits.

VI.

SPACE AND HYPERSPACE.

BY "TESSERACT" (CLAUDE BRAGDON, ROCHESTER, N. Y.).

The baffling thing about speculation concerning the fourth dimension of space lies in the fact that we must reshape our very idea of space. We naturally think of space as the box which contains all the furniture of consciousness, and in altering our conception of it, as it is necessary to alter it in postulating an additional dimension, we are dealing, not with the contents, but with the box.

Let us think, not of space, but of *spaces*, differentiated from one another by their dimensionality and designated in terms of it, so that the greater the number of its dimensions the "higher" will be the space. Let us think of each higher space as generated from the one next below it, and as having the properties and dimensionalities of all spaces lower than itself patent, and higher than itself, latent.

Our space has three dimensions, and within it are given the conceptions of point and line, line and plane, plane and solid. These involve the relation of our space to higher space, and of lower space to our own.

One segment of a straight line is separated from another by a point, and the straight line itself can be generated by the motion of a point. One portion of a plane (2 space) is separated from another by a straight line, and the plane itself can be generated by the movement of the straight line in a direction not contained within itself. Again, two portions of a solid

(3 space) are separated from one another by a plane, and the plane, moving in a direction not contained within itself, can generate the solid. From this it is possible to formulate a definition of space irrespective of its dimensionality: *Space is that which separates two portions of higher space from each other.* Also: *Our space will generate higher space* (i. e., 3 space will generate 4 space) *by moving in a direction not contained within itself.*

In the generation of the plane by the line, and the solid by the plane, the "direction not contained within itself" is inevitably a direction at right angles to the line and to every line of the plane. Hence, *a movement in the fourth dimension is a movement in an unknown direction at right angles to every known direction embraced within three-dimensional space.*

Proceeding now from general to particular, let us endeavor to form some idea of the simplest symmetrical four-dimensional solid — a tesseract — corresponding with a square in 2 space and a cube in 3 space.

In 2 space a square surrounded by four other squares, one on each of its four lines, would be completely bounded and inclosed; but if this same square, together with its surrounding squares, moved in a direction at right angles to its surface (i. e., out of 2 space into 3 space) a distance equal to the length of one of its sides, it would trace out a cube bounded by four other cubes. To inclose it completely in 3 space it would be necessary to add two more bounding cubes, the first to that face which coincides with the square in its first position, and the second with the square in its final position, i. e., in the positive and negative ways of the third dimension. The cube would then be completely bounded and inclosed in 3 space. Imagine now that the cube, together with its six surrounding cubes, moved in a direction at right angles to its every dimension (i. e., out of 3 space into 4 space) a distance equal to the length of one of its edges, then it would trace out a higher cube, or tesseract, and each of the six surrounding cubes, carried on in the same motion, would trace tesseracts also, grouped around the original center tesseract. But would they inclose it completely? No; because as in the former case there would be nothing between the cube and that from which its motion started. The movement in the new dimension would not be bounded by any of the six cubes, nor by what they formed when moved. It would therefore be necessary to add two more bounding tesseracts, in the positive and negative ways of the unknown, or fourth dimension, at the beginning and at the end of the motion.

In this manner it is established that a tesseract is completely inclosed by eight similar tesseracts; and because the faces of a tesseract are cubes, *a tesseract is bounded by eight equal cubes.*

Now just as the cube has squares, lines, and points as elements, so the tesseract has cubes, squares, lines, and points as elements. Let us examine these.

In the movement of a cube, which consists of six squares, twelve lines, eight points, into 4 space, the six squares would give six squares in their initial, and six in their final position; and each of the twelve lines of the cube would trace out a square. Hence, *a tesseract is bounded by twenty-four equal squares* $(6 + 6 + 12)$, and further analysis by means of models or diagrams reveals the fact that *each is a meeting surface of two of the cubic sides.*

The twelve lines of the cube, in its movement into 4 space, give twelve lines of the tesseract in their initial, and twelve in their final position, while each of the eight points traces out a line. Hence, *a tesseract is bounded by thirty-two lines* (12 + 12 + 8), and further analysis by means of models or diagrams reveals the fact that *each is common to three cubes or to three square faces.*

The eight points of a cube, in its movement into 4 space, give eight points in their initial, and eight in their final position. Hence, *a tesseract has sixteen points* (8 + 8), and further analysis shows that *each is common to six square faces and to four cubes.*

Although by these means it is possible to form a conception of the elements and projections of a tesseract in our space, and even to depict them graphically by a series of related diagrams, the intellect fails in its effort to co-ordinate these into one figure, that is, to picture the tesseract itself. The chief difficulty lies in the fact that it is next to impossible to think of a cube, a solid of our space, as a mere boundary—one of the sides of a higher solid. A study of the corresponding predicament as presented to 2 space consciousness will be of assistance here.

In a hypothetical plane-world, to a hypothetical plane-being, endowed with a body and a mind like our own, but minus the power of movement in the third dimension and, therefore, minus the consciousness of it, a square would be a solid body, being completely enclosed by boundaries, in the form of lines, through which he can neither see nor pass. The essential insubstantiality of such a body and its property, known to us, of being one of the boundaries of a solid in our space, would seem to him no less a paradox than the cube as a mere boundary of the tesseract seems a paradox to us. The square rests in 2 space, and to the consciousness of that space it *is* a solid if we define a solid as a completely bounded figure the interior of which cannot be reached without the disturbance of its boundaries. According to the same argument, a cube in our space is a solid only to our perception, and with relation to our space. In 4 space, or to four-dimensional consciousness, it loses its "solidity" in becoming the boundary of a higher solid, for *the solid of any space becomes the boundary of a corresponding solid in higher space.*

A rotation in 2 space takes place about a point; in 3 space about a line; hence, by analogy, *a rotation in four dimensions takes place about a plane.*

In 2 space, right-handed and left-handed similar right-angled triangles could never be made to coincide by any motion proper to that space, but their perfect coincidence could be effected easily by the rotation of one of them in the third dimension, about the line of one of its sides. So, in our space, corresponding right and left-handed solids of the same elements and equal volume, like the right and left hands, for instance, could be made to coincide by the rotation of one of them about a plane. The mirror image of a solid represents the solid after such a rotation.

The number and the variety of deductions concerning 4 space which can be made from simple premises of the above order is almost infinite, but a sufficient number of examples have been given to explain the *what* and *how* of 4 space. *Where* is it?

Go back to our first definition: *Space is that which separates two portions of higher space from each other.* Conceive of 2 space therefore as a vertical plane, separating two portions of 3 space from each other. Now, in order that this separation should be effective, the plane must be something more than a mere geometrical abstraction, that is, if it is a "real" plane, it must have a very slight thickness. Its particles will have a free movement and circulation in the two principal dimensions of the plane, but their power of movement in the third dimension, being limited by its thickness, which we assume to be so slight as to be inappreciable, will be confined to the infinitely minute. This is the hypothetical space of the hypothetical two-dimensional "man," but if he were set down in it, without some world to tread, some solid ground to push off from, he would be in a condition analogous to that in which we should be if we were suspended free in space. Let us give him his world: this would naturally be a vertical disk, the cross-section of a sphere, made of the matter of his space, held together by an attractive force analogous to gravity, which not only makes and preserves the form of his disk world, but holds him to the rim which is its surface. The direction of this attractive force of his matter would give him a knowledge of up and down, determining for him one direction in his plane space; also, since he can move along the surface of his earth, he will have the sense of a direction parallel to its surface, i. e., forward and backward, but he will have no sense of right and left, the direction extending out into our space, which is his higher space. This would be for him the unknown dimension. With the first step in the apprehension of 3 space he would come to the conviction that if the third dimension exists, the objects of his world which he had conceived of as geometrical figures of two dimensions only, had a certain, though a very small, thickness in the third dimension, that the conditions of his existence demanded the supposition for an extended sheet of matter, from contact with which in their motion his objects never diverged.

Exactly analogous suppositions must be formed by us with regard to 4 space, namely: that our space separates two portions of higher space from each other; that in the infinitely minute of our world there is extension and the power of motion in the fourth dimension; that there is a direction toward which we can never point extending from every point of our space, and that we "slip along" this invisible wall of higher space which we must give up any attempt to picture in relation to ours just as a plane being would have to give up any attempt to picture the plane at right angles to his plane.

Kant imagined that space might contain more than three dimensions. He even infers their "very probable real existence." Gauss and the non-Euclidean geometers have established a distinction between laws of space and laws of matter which clears the way for a conception of space of any dimensionality. To such a conception mathematics lends itself in a truly remarkable manner. It is reasonable to suppose that the fourth power of a number should have its spatial equivalent, just as a square is the spatial equivalent of its second power, and a cube of its third. Moreover, it is just as possible to deal with four dimensions arithmetically as with three and by analogous operations, and the shapes, movements, and mechanics of simple four-dimensional solids can be made intelligible to the understanding—in other words, the mind finds itself still at home in regions where the senses do not operate.

The fact that we can apprehend but three dimensions does not disprove the existence of a fourth, and for the following reason. All our strictly sense impressions are two-dimensional, for we can see and contact only surfaces. Touch teaches that an object retains the same form and extension through all the variations of distance and position under which it is observed, notwithstanding that the form and extension of the image on the retina change constantly with the variation in position and distance of the object in respect to the eye. The reconciliation of the apparently contradictory facts of the *invariableness* of the object and the variableness of its appearance is only possible in a space of three dimensions, in which, owing to perspective distortions and changes, these variations of projection can be reconciled with the consistency of the form of a body. Consequently we come to the idea of the third dimension by an intellectual process in order to overcome the apparent inconsistency of facts of the existence of which our experience daily convinces us. This being so, the moment we observe in three-dimensional space contradictory facts, our reason would at once be forced to reconcile these contradictions, and in that attempt a conception of a fourth dimension of space—if it reconciled the contradiction—would arise. Furthermore, if from our childhood phenomena had been of daily occurrence, requiring a space of four dimensions for their proper understanding, we would naturally grow up with the conception of a space of four dimensions. It follows that the *real* existence of 4 space can only be decided by an observation of facts.

Are there any facts? Many phenomena classed as "occult," clairvoyance, apparition at a distance, the moving of ponderable objects by unseen means, etc., can be explained, on their mechanical side, on the theory of a fourth dimension; but as the dispute as to the reality of these phenomena is still going on, the reality of the fourth dimension may be said to be an open question.

VII.

AN INTERPRETATION OF THE FOURTH DIMENSION.

BY "QUEFANON" (ARTHUR HAAS, NEW YORK CITY).

A ship in a canal could be located at any given time by a knowledge of its distance from some town, since its motion from that town has been restricted to one direction. When space is of such a nature that a point in it may be located by one measurement from some fixed or standard point, that space is said to be linear or one-dimensional.

The same boat on the ocean, however, could not be located unless two measurements were given—its latitude and longitude. The nature of *such* space is defined by the words "surface" or "two-dimensional area."

If, now, our vessel were converted into an airship or a submarine, we should be obliged to add to our other data its distance above or below the sea level in order to place it accurately. With three basic elements (in our illustration; the equator, the prime meridian, and the sea level) and with three known distances from these elements, we can locate any point that comes within our consciousness, whether above, on, or below the surface of the earth. Any additional measurements would be either superfluous or misleading. Hence we say that our space is three-dimensional.

In this discussion it will be necessary for us to use graphic representations of changes in one-dimensional, two-dimensional, and three-dimensional space, and for this purpose we shall adopt as illustrations respectively the movement of mercury in a common thermometer, the movement of the arms of a semaphore, and the physical changes which a jellyfish undergoes in the course of its development.

The rising and falling of the mercury is a one-dimensional movement. If we wish to keep an automatic record of the temperature during a given period, it would be an easy matter to pass a strip of photographic paper behind a thermometer, and allow the sun or some artificial light to darken the part above the mercury. If this paper were kept stationery, the only record we could obtain would be that of the minimum height of the mercury. Therefore, some movement of the strip is necessary. If this motion were to be in the length direction of the thermometer, every part of the paper would be exposed to the action of the light, and no record at all would be obtained. We could obviate this trouble, however, by covering the strip while it moved through a distance equal to the length of the thermometer, then exposing it for a short time, and then again moving it. Thus, without involving a second dirnension, we would get a permanent record of various successive heights of the mercury. These pictures would be intermittent, and we would miss the changes that took place while the picture film was moving. In order to get a complete and continuous chart of the changes, we must move the paper in a direction other than that of the length of the thermometer. In short, we are forced to introduce a second dimension. The strip may be moved by clock work, and then we would have a two-dimensional chart, from which we could determine the temperature at any required time, the horizontal measurement showing the time of observation, and the vertical one the height of the mercury at that time. The result of this experiment could be read by passing this chart behind a vertically slotted surface, thus obtaining the effect of a line whose length varies as the strip of paper slowly passes the open space. These variations will, of course, exactly reproduce the variations in the height of the mercury.

It is not difficult to imagine a being whose percepts are confined to a linear representation of objects; for instance, a man whose sense of touch is paralyzed and whose eye is covered by a cataract in which a vertical slit has been successfully cut. Better yet, we may conceive of one whose retina itself is merely a line instead of a spherical surface. He could not imagine such a thing as an angle, and it would be as hard to explain parallel lines to him as to describe color to a man born blind. He could see the changes in the height of the mercury just as well as we, but a triangle passed before his line of vision would present the same sort of picture, viz., a line increasing in length; and there would be no way of convincing him of the simultaneous existence *of all its parallel elements,* which to us is a very simple concept. He could, however, picture from his memory, and re-produce, two or more lines which represent the height of the mercury at different times, but they would all lie in his one-dimensional consciousness as separate pictures.

His knowledge of a growing tree would be confined to a line with various colored parts which change, both as the tree grows and as he moves his line of vision, but the most complex of these changes could be reproduced by a picture on a plane surface, slowly passed before his eye. In brief, such a being could

have a *perfect conception of one-dimensional change*
merely through a two-dimensional representation.

When we come to consider changes in two dimensions, such, for instance, as are caused by the motion
of the arms of a semaphore, how are we to represent
them. A series of photographs might be taken in rapid
succession, and if these were placed behind each other,
a solid would be formed of which we might say each
picture was a cross section. A book made up of these
pictures in their order is such a solid, and the little
pocket mutoscope exactly satisfies this description. If
its pages are rapidly turned, the successive sections
are presented to our sight, and we apparently see the
arms of the semaphore changing their position. The
kinetoscope with its two-dimensional strip and its
shutter does the same thing more steadily, and presents
the illusion of motion in a two-dimensional area even
better than the little hand mutoscope. The pictures
taken by the mutograph are really always two-dimensional; it is only our experience in shadow and perspective which gives us the illusion of motion in three
dimensions when the ordinary "moving picture" is
thrown on the screen. If we left the camera film
unmoved while the semaphore was moving, only a
picture of the stationary parts would be taken, the
rest would be a blur. Hence we must move our picture
film.

If we move it continuously, no record of any position of the semaphore will be taken. Here again we
must obviate the difficulty by shutting out the light
while the film moves over a distance equal to the size
of the picture it is to take, then exposing it, and then
covering it again. But no matter how quickly the
camera shutter is snapped, the representations of the

mutograph can never be continuous. In order to represent continuous and gradual change from one position of the semaphore to another, a line must be used for every point in the semaphore arms, and this line cannot usually be represented in the same plane as that in which the motion takes place, without interfering with the path of some other point in the moving object. A new dimension must be introduced to make a record of a really continuous change. Thus, a more nearly correct, though much more difficult, method of physically representing the phases of the semaphore arms would be the following: Suppose a plastic material (like wax) to be forced against the semaphore while its arms are moving. A continuous opening would be left in this material as the semaphore is forced deeper and deeper into it. Suppose again that this opening were filled with plaster of Paris, and that the wax were melted away. We would then have left a solid body, *every* section of which would represent a phase of the semaphore, and which would contain in itself *every* position that the movable arms had assumed during the course of the experiment. This representation is in what we ordinarily call the solid form; that is, three-dimensional.

If an imaginary being with a two-dimensional sense, an "Inhabitant of Flat-Land," were to have this solid passed through his plane, he would see reproduced the continuous motion of the semaphore arms. Like our slit-eyed friend, the "Line-lander," and for analogous reasons, he could not conceive the simultaneous existence of all these cross sections. But by using his memory, he could reproduce some of them as separate pictures in his two-dimensional world—such pictures, perhaps, as we have in our kinetoscope film.

If a small quantity of yeast were allowed to ferment between the slide and cover glass of a microscope, we should have under our observation the growth of an object in practically two dimensions. Now, its phases at very small intervals could be photographed, but the same conditions that met us in the case of the semaphore, face us again. The only way to represent all the changes that take place would involve the tracing of each point from one position to another. This would produce a line; and since two dimensions are required to present all the points in their relative positions at any given time, this line, in order not to be obscured, must extend beyond the two-dimensional space in which the growth takes place. We must, therefore, create a solid, whose successive sections would be recognized by the two-dimensional mind as the growth of the object which was passing through the plane of their consciousness.

In our previous illustrations we were able by the use of *two-dimensional* space to fix permanently variations of position and magnitude of a *one-dimensional object,* and in *three-dimensional space* we were able to fix permanently the changes of an object moving or growing in two dimensions.

Coming now to the phenomena of our every-day world, we know that changes in position and growth take place continuously in our three-dimensional space, and that the time element is necessary to determine exactly the conditions of any variable or movable thing. Thus the description of a tree would give an entirely false impression, if only its dimensions were given without adding the particular time when these were taken; and the position of a planet would be incompletely given, unless the time of observation were reported together with the other three necessary measurements; even as the position of a ship upon the earth's surface is not known by its latitude and longitude unless we know also when these were calculated, and the idea of the temperature of a body would be incomplete unless the record of time accompanied the statement of the mercury's height above the zero mark.

If we could only picture to ourselves that a three-dimensional object is merely the *cross section* of a permanent four dimensional thing, that what we are cognizant of is merely a *phase* of a thing which exists in its entirety, and of whose other phases we are ignorant, till they are brought to our own consciousness or till our consciousness reaches them, then we could conceive the physical nature of a four-dimensional object.   Considering, for instance, our own material bodies, we are conscious of a gradual change of shape and position of all the parts, and yet, at the same time, we are conscious of a continuing identity throughout all these changes.   Our past experiences are as real as the experiences we are now undergoing.   Those past experiences, or phases of our existence, are as much a part of us as the present ones, and yet owing to the limitations of our three-dimensional consciousness we can reproduce past conditions only in memory.   Nevertheless our lives in their completeness are made up of the sum of *all* our experiences and if our *whole* lives are considered as units, and each period of which we are conscious requires a three-dimensional space, then each individual may be considered as a four-dimensional solid.

Let us, however, take a more simple illustration. A biologist wishes to present to his class a concrete means of studying the jellyfish. He orders his pattern-maker to model perhaps fifty copies of the animal in question, showing the changes from the egg to the perfect adult. These are molded in glass, and are brought into the classroom for study.

Now, although every particle of the living jellyfish is constantly changing, either in size, or position, or in its relation to neighboring particles, we say it is the same jellyfish; there is a something that persists through all the changes; an individuality which differentiates this animal from all others, although to-day it is as different from what it was previously as any two models.

These models may be considered copies of mere phases of the jellyfish, just as photographs may be said to represent phases of the fermenting yeast, and two separate lines may be said to represent corresponding phases of the mercury length in the thermometer.

But no matter how small the interval which elapses between the making of two successive models, if there be any change at all, that change must have involved many, nay an infinite number, of smaller changes, and these changes in the case of each atom of the living organism must have been continuous; that is, they *must be represented by a line,* and not *by a succession of separated points,* if we would preserve the individuality of the animal in question.

Now this line cannot be represented in our three-dimensional space without interfering with other atoms which surround it in three directions. We are compelled, therefore, as in the previous illustrations, to go outside the space in which the change takes place, in order to represent completely the continuous change in anything which preserves its individuality while changing. Hence, to represent graphically a gradual change or growth in a three-dimensional object, a four-dimensional space is necessary; and the representation in such space of a fixed and permanent object which combines all the phases of a three-dimensional solid would constitute a four-dimensional figure.

Mind you, I do not say that a growing jellyfish is necessarily a fixed four-dimensional object, passing through three-dimensional space, but I do say it could be so represented; and that then a four-dimensional mentality could see any or all of its three-dimensional phases simultaneously, just as we can in a two-dimensional chart perceive simultaneously all the lengths of a varying line. To get a vague conception of such a four-dimensional figure, it is necessary for us to group all our three-dimensional memories of some changing object between two definite times, and imagine them merged into a something of such a nature that no part of one memory picture overlaps a different part of another, and yet that each of these concepts is itself complete. This is, of course, impossible to most of us, but so are many other mathematical and physical concepts.

More scientific but somewhat similar considerations than those quoted above, have forced all the great mathematicians and many great physicists to accept the fourth dimension as a solution of many difficulties. Its use is recognized, almost unconsciously, even by the elementary student, when he computes the area of a triangle, for here he multiplies four dimensions and extracts their square root to obtain a two-dimensional result, namely, $\sqrt{s(s-a)(s-b)(s-c)}$. Furthermore, this theory lends itself to the simplification of many physical and metaphysical problems. Therefore, its adherents find an ever-increasing army of converts.

At present our three-dimensional knowledge is itself very imperfect. We can move unrestrictedly in two dimensions, but when we attempt to travel in the third, we are limited more than the fishes or the birds. Our knowledge of the interior of solids is so dependent upon surface study, that in order to scientifically study a single cubic inch of tissue, we must examine thirty thousand square inch sections cut by a very fine slicing machine (the microtome).

The *transparency* of the jellyfish was the exceptional feature which permitted its use to illustrate a three-dimensional object whose changes could be studied without dissecting it.

Our three-dimensional concepts generally are mere inferences from our two-dimensional knowledge, and we are easily deluded by our senses in forming them. When our knowledge of solids becomes as nearly perfect as our present knowledge of surfaces, then the vague four-dimensional figure may assume a more concrete form. Will this ever happen? Who can tell? Many more revolutionary theories have found concrete expression, and then obtained a firm foothold against stronger opposition and with less necessity for their existence.

VIII.

LENGTH, BREADTH, THICKNESS, AND THEN WHAT?

BY "QUESNEL" (LEONARD C. GUNNELL, SMITHSONIAN INSTITUTION, WASHINGTON, D. C.).

It is difficult for a finite mind to picture, or even to conceive of conditions unconnected with finite experiences, and unperceivable by finite senses. All finite experiences are connected in some way with material substances or with perceivable forces. All material substances have one or more of the properties of length, breadth, or thickness, and all physical forces may in some way be rendered perceivable.

To the lay mind many scientific achievements seem almost miraculous, though by systematic effort any educated mind may comprehend any of the achievements in any of the sciences, for the results have to do alone with matter and forces, and are expressed in terms which may be transposed into the equivalent terms commonly used to describe every-day actions and experiences.

The science of astronomy, dealing as it does with infinite masses, infinite forces, and infinite distances, would seem to require the ultimate effort of a finite mind to comprehend, but the ultimate problems in astronomy deal only with masses, forces, and three-dimensional space, things of common knowledge, connected only in a lesser degree with common every-day actions and experiences.

The ultimate theories in physics and chemistry deal with atomic and molecular forces and masses and with their interactions; no matter how vast or how minute are the masses or forces they remain masses and forces, and their dimensions and activities are described in terms equivalent to those used in describing all other qualities and actions.

The qualities of three-dimensional matter we comprehend, forces we comprehend, and vibrations we can comprehend as one of the manifestations of forces; consequently, when the chemists or physicists in dealing with ultimate theories claim, as they do, that matter is simply the manifestation of forces, the idea may be grasped, though it may or may not be accepted.

Advocates of the fourth dimension ask more of our reasoning powers in explaining their hypothesis. One must lay aside all usual comparison with concrete things in grasping this hypothetical idea, as we can only reason about the qualities possessed by such a transcendental figure, the exact nature and form of which cannot possibly be definitely pictured to a finite mind. The conception is mental purely and is not connected with, nor necessary to, the solving or understanding of any actual problem. Four-dimensional space is not and cannot be connected with finite problems or experiences limited, as all such problems and experiences are, to space of three dimensions. We live and exist in space, all our problems and experiences are limited to actions in space.

We know that a point has position alone, but position in space with no dimension; when the point moves in a straight path a line is traced which has length alone, the first dimension, beginning at a point and ending at a point. Should the line move at an angle, say at a

right angle with itself, a plane is formed having two
dimensions, length and breadth, with a line at the
beginning and a line at the ending of its path, and in
addition two new lines traced by the two points in their
movements.   If the motion of the line is at right angles
with the path of the point and for a distance equal to
the length of the line a square is formed in a plane.
A square being a good representative two-dimensional
plane figure, we will use it in our explanation.   In
other words, a plane square is a figure having length
and breadth, is bounded by four lines of equal length
which meet at four points.   In a similar manner a
cube is formed by moving the plane, at a right angle,
a distance equal to the length of the line; this cube will
have thickness, the third dimension, in addition to the
length and breadth of the line and the square.   As it
begins with and ends with a square and each of the
four lines bounding the first square will, by its movement, trace a new square, it will be bounded by six
squares.   It will also have four lines from the original square, four lines in the final square, and four
lines traced by the movements of the four points of the
original square, or twelve lines in all, meeting at eight
points; four points from the original square and four
points from the final square.

Let us assemble the above facts for convenience in
comparing them and add to the table the corresponding properties of an imaginary fourth dimensional
figure, these being determined as follows:

As the line, the first dimension, is formed from a
moving point, so a square, a typical second dimension
figure, is formed from a moving line, making a figure
bounded by four lines, and as a cube having a third
dimension is similarly formed by a plane moving into

	Number of dimensions in figure	Number of points in figure	Number of lines bounding figure	Number of planes bounding figure	Number of cubes bounding figure
Point..........	0	1	0	0	0
Line...........	1	2	0	0	0
Square	2	4	4	0	0
Cube..	3	8	12	6	0
Corresponding figure of four dimensions....	4	16	32	24	8

the third dimension, making a figure bounded by six planes, does it not follow that a corresponding fourth dimensional figure is formed by the movement of a cube into the fourth direction and will be bounded by cubes?

If this is the case and the line derives from the point two points, and the square derives from the line four lines and four points; and if the cube derives from the square eight points, twelve lines and six planes, does it not follow that the moving figure gives to the corresponding fourth dimensional figure the following qualities?

The cube at rest has eight points in space, at the end of its movement it has eight new points in space, its movement into the fourth dimension has created the fourth dimensional figure; therefore, the figure should have sixteen points. The cube has at rest twelve lines or edges and has at the end of its movement twelve additional lines, and each of its eight points has traced a new line, making thirty-two lines or edges in all for a corresponding fourth dimensional figure. Similarly, as the cube has six planes at the beginning and has six new planes at the ending of its movement, and as its twelve lines will in moving trace twelve new planes, there will be twenty-four planes in the fourth dimensional figure. Now, as a cube is generated from a moving square, when the cube moves to generate a figure of the fourth dimension, the new figure will have a cube at the beginning of the movement and another cube at the end, and in addition each of the six squares bounding the original cube will by their movement trace a new cube, thus adding six new cubes to the two already mentioned, or eight cubes in all to bound the new fourth-dimensional figure.

From this line of reasoning we derive from a point in an ascending scale through the well known figures and attributes of the first-, second- and third-dimensional figures, the logical attributes of a hypothetical figure of four dimensions, which is that it is bounded by eight cubes and has twenty-four planes and thirty-two lines meeting at sixteen points.

It is not sufficient to say that the incomprehensible fourth-dimension of geometry, corresponding to the figure of the fourth power of arithmetic and algebra, does not exist because we cannot picture it or even conceive of it, or because it does not enter into any problem connected with known matter or force. It may properly be claimed that a three-dimensional figure of infinite length, infinite breadth, and infinite thickness would embrace infinite space; but is it possible to picture or comprehend what infinite space is? Can a finite mind picture a space with no ending, space with no beginning and no ending; limitless space in which our vast solar system is a mere dot, in which the known stellar universe is probably also comparatively a mere dot, although it is actually so vast in extent that the light from some of its component stars which started toward us generations ago or centuries ago is only now reaching us. All this known space is, however, so far as the human mind can picture it, three-dimensional, though its vastness is well nigh incomprehensible.

If space is limitless, the idea is incomprehensible, and if it is limited its limits are incomprehensible. Space is limited or it is limitless; in either case the idea is incomprehensible. Thus the mere statement that an idea is incomprehensible does not prove its non-existence. It is common to use as an analogy, in explaining the idea of the fourth dimension, the possible experiences of hypothetical beings existing in space of more limited dimensions than the three dimensional space we understand, and thus by comparison picture our possible experiences with space of four dimensions. Picture a being whose existence is passed in a plane, say a finite two-dimensional figure, a square, for instance. This being would be shut in by the four lines bounding the square, there would be no upper side or under side imaginable to this being, for upper side or under side would imply thickness which would be a dimension higher than the plane. Now this being could move in any direction on its square until a boundary line was encountered, which would be to it a barrier; it could picture the other side of this line, for the other would be simply a continuation of the plane; but to reach the other side without passing through the line would be incomprehensible, for it would necessitate movement in the third dimension, a movement in a direction incomprehensible to the plane being.

Now, however, a three-dimensional being, able to move and act in three-dimensional space, a human being, for instance, could remove the two-dimensional being from its square, pass it over a boundary line and back on its plane outside of the boundary lines of the square. Thus the two-dimensional being would find itself on the outside of its barriers without having passed through any of them, for its movement in the third dimension would have been unperceived and incomprehensible. Now, imagine a being in a cubic three-dimensional figure, say a box having solid covers on all of its six sides. There is no conceivable way of getting out of such a box save by passing through one of the six sides, yet from the analogy derived from the experience of the two-dimensional being a fourth-dimensional being could move the being confined in the box into the fourth dimension, and so out of the box without passing the being through the sides of the box. This act is no more incomprehensible to the human three-dimensional being than would be the act of passing over the boundary line to the two-dimensional being.

It is obvious that a one-dimensional figure on a line can, by motion in the second direction, pass off of the line without passing through the points which begin and end the line, and we have shown that a two-dimensional figure can, by motion in the third direction, pass out of a square without passing through the square's boundary lines; therefore, a three-dimensional figure could, by motion in the fourth direction, pass out of a cube without passing through the cube's boundary planes.

It will be noted that the generation of each of the three figures of known space is accomplished by one of three distinct motions, each differing in direction from the motions preceding, and that by one or a combination of these three motions any point of any conceivable figure of known space can be reached. Now, therefore, this fourth movement, that is, the movement of the cubic figure, in generating the fourth dimensional figure, is a movement differing essentially in direction from the movement the plane makes in generating the cubic figure, just as the line movement in generating the plane differs essentially in direction from the movement the point makes in generating the line.

The fourth movement, essential to the generation of a fourth dimension from a third-dimensional figure, is inconceivable to the human three-dimensional being, just as the third movement essential in generating a third dimension from a two-dimensional figure would be inconceivable to a two-dimensional being whose possible experiences were always limited to a plane.

It is not logical to state that a fourth dimension cannot exist, for from the analogies derived from the other three movements, the first that of a moving point generating a line, the second that of a moving line generating a plane, and the third that of a moving plane generating a cubic figure, a clear strong argument is derived for the possibility of a fourth movement differing essentially in direction from any of the three preceding movements or any combination of them, just as they severally differ essentially from each other. This fourth movement is the movement necessary to generate a fourth dimension whose figure is inconceivable to the finite human mind, but whose boundaries, qualities, and other attributes can be as definitely described as if the hypothetical figure could be perceived by the human senses of vision and touch.

IX.

THE FOURTH DIMENSION ALGEBRAIC-
ALLY CONSIDERED.

BY "N." (BURTON HOWARD CAMP,
MIDDLETOWN, CONN.)

The concept of the fourth dimension is exclusively a mathematical one, and, therefore, can hardly be made intelligible without the introduction of a few mathematical ideas. The more important aspects of it, however, I shall endeavor to explain with the use only of the elements of that algebra and geometry which are usually taught in high schools.

The reader will recognize the following as types of equations with which he has dealt, though he may not recollect clearly all their properties:

$$x + y = 4 \qquad (1)$$
$$x^2 + y^2 = 1 \qquad (2)$$
$$2x^2 + 3y^2 = 1 \qquad (3)$$

Here the letters x and y are, in first courses in algebra, commonly called the "unknowns." I do not propose to inquire what values these unknowns may have, and, of course, these equations are not supposed to be true simultaneously; they are chosen almost at random as three entirely separate and independent examples to illustrate the fact that some equations contain two and only two unknowns. In other equations we may put three unknowns; in still others four, or five, or as many as we like.

$$x + y + z = 4 \qquad (4)$$
$$\text{and } x^2 + y^2 + z^2 = 1 \qquad (5)$$

are examples of equations in which the number of
unknowns is three, and they are x, y, and z.

$$W + x + y + z = 4 \qquad (6)$$
$$\text{and } W^2 + x^2 + y^2 + z^2 = 1 \qquad (7)$$

are equations in which their number is four.

Now, just as illustrations are valuable in making
language vivid, so the mathematician finds that, when
he can form some sort of a picture of his algebraic
work, he realizes more clearly what it means; and it
happens, fortunately, that there are a number of ways
in which he can form pictures of such equations as
these. I shall speak of but one, the simplest and common method. According to this method, in order to
form pictures of equations in *two* unknowns, like (1),
(2), and (3), it is necessary to use space of two dimensions—for example, a plane; the essential thing is that
anywhere in this space it must be possible to conceive
of two lines intersecting at right angles. Some readers will recognize this as the method of rectangular
Cartesian co-ordinates, but it is not important that the
principle be explained in detail, for all we shall need to
know is that it exists. It turns out that the picture we
get for equation (1) is a straight line, drawn, of
course, in some plane; that equation (2) is a circle,
and that equation (3) is an ellipse; there are besides a
host of other curves, corresponding to all conceivable
algebraic equations in two unknowns—spirals, heart-
shaped curves, "figure eights," etc., some of which
have been given names, and some of which have not.

In order to represent equations like (4) and (5),
in which the number of unknowns is *three,* space of
two dimensions will not suffice; now we shall need to
let three straight lines intersect so that each makes a
right angle with the other two, and that cannot happen

in space of two dimensions. Three adjacent edges of a cube, however, are known to be mutually perpendicular, and so the ordinary space of three dimensions to which we are accustomed will be suitable, and by its aid we will be able to picture these equations. It happens that the representations of equations (4) and (5) will then be surfaces; equation (4) will be a plane surface, and equation (5) the surface of a sphere; and here again we may write any number of equations in three unknowns, and each will be representable by some surface—perhaps plane, perhaps gently curving, perhaps full of convolutions so that it folds in and out upon itself.

When, therefore, an equation contains exactly two unknowns or exactly three unknowns, it can be represented thus by some curve drawn in space of two dimensions, or by a surface in space of three dimensions. But when the number of unknowns is increased to *four*, as in equations (6) and (7), the method fails; for now it requires a kind of space in which may be drawn four straight lines, all meeting at one point, and each perpendicular to the other three. It is not possible to conceive of such a situation, and, therefore, the mathematician is obliged to do without the representation he has thus naturally been led to desire. But, though he cannot have the picture, he can have the language. Equation (6) looks a good deal like equation (4), which is a plane, and indeed it has many of the same properties; so he decides to call (6) also a plane, but to distinguish it from (4) he calls it a "plane in four dimensions,"* while (4) is a plane in three dimen-

---

* These are not suitable terms, for an actual plane or a sphere may be spoken of as in space of four dimensions. "Hyperplane" and "hyperspace" are terms often used.—H. P. M.

sions. Likewise (7) is to be called a "spherical surface in four dimensions," while its analogy, (5), is a spherical surface in three dimensions. He does not mean to imply by such language that it may be possible to conceive of four mutually perpendicular straight lines; he does not suggest anything whatever about our ideas of space, or, to speak more precisely, about our ideas of motion. He is merely using analogous terms because he finds them convenient. They possess for him some valuable qualities—they are brief and suggestive; and so, with full knowledge of their limitations, he uses them. They are brief, because it is generally shorter to give merely the name of a surface, than it is to describe minutely the general class of equations which that surface represents. It would take us too far afield to show fully in what ways he finds them suggestive, but a single illustration may be helpful. Suppose he wishes to find out what relations exist between all equations which, like (6), he has decided to call planes in space of four dimensions, and all equations which, like (7), he has decided to call spherical surfaces in space of four dimensions. These are equations in four unknowns; he looks back at their analogues in three unknowns, that is, at equations like (4) and (5), for which he has really found a geometrical meaning. These are really representable by the plane and by the spherical surface; and so, by thinking of the geometrical relations between these two figures, he has a clue to what he is to look for in dealing with the corresponding equations in four unknowns. Of course, he may not find it, for it is not true that always the same relations hold for these different sets of equations, but at least he is on the road to discovery—if he does not find what he is looking for, he is liable to find something else.

From this point of view, then, the fourth dimension is a convenient phraseology, and only that. It is customary also to use in like manner the terms, "fifth dimension," and "sixth dimension," and so on, in speaking of equations in more than four unknowns; and when the mathematician thus uses such terms, when, for example, he speaks of a surface in four-dimensional space, he is speaking and thinking merely of some kind of equation in four unknowns.

But there is another point of view from which the fourth dimension is sometimes considered. Hopeless as it is for us, who have lived only in three-dimensional space, to conceive of four straight lines meeting at a point so that each is perpendicular to the other three, yet it is quite possible for us to find out what sort of things would happen if indeed four such straight lines could exist. To assume, then, that four such straight lines may exist, and to deduce the logical results of that assumption is another of the mathematician's problems. It matters not to him that his assumption asserts an inconceivable situation; he is not concerned at all with the question of its truth, only with its logical consequences.

Of course, such a geometry does not at the present state of our knowledge have important practical applications, but at least it is rich in ideas, and it is by no means certain that its relation to our surroundings is not closer than it appears. For, though in this sense four-dimensional space, that is, motion in four different mutually perpendicular directions, is to us unthinkable, we cannot surely say that it may not exist. If it does exist, we can know something of those four-dimensional bodies which may also exist, and a number of interesting results follow. Suppose, for example, we consider some closed two-dimensional figure, say a circle. We know it is impossible for a point which always remains in the plane of the circle to move from that part of the plane which is inside the circle to that part which is outside, without passing through the circumference. But, if the point may make use of motion in a third dimension, and so get out of the plane for an instant, it may jump over the circumference, and without touching it at all reach the outer part of the plane. Likewise, if we try to think of a point moving from the inside of a sphere to the outside, without passing through the surface, the thing is inconceivable to us, and so we say it is impossible; but, if we assume a fourth dimension, then the point could, so to speak, "jump over" the surface, and appear again in three-dimensional space outside the sphere. The same is true of any such closed surface in three dimensions. If a prisoner could make use of motion in a fourth dimension, we know he could escape from the inside of a closed cell without touching the sides at all.

From these two aspects, then, the mathematician commonly regards this subject of four dimensions— one furnishes an abbreviated and suggestive method of denoting various types of equations in four unknowns, and the other is the supposition that four mutually perpendicular straight lines can exist. Neither can properly be the basis of any physical theory, at least at present, for the one is only a phrase and the other is a supposition which is not surely supported by anything that we know of the physical universe. At the same time, it may be well to remember that there is nothing self-contradictory in the assertion that each of four straight lines can be perpendicular to all of the other three. Whatever "proofs" have been given that this is impossible are based (ultimately) upon the intuition that space is three-dimensional. In other words, the only reason we have for believing that only three straight lines can be mutually perpendicular is that such a condition is the only one we have ever experienced.

X.

DIFFICULTIES IN IMAGINING THE FOURTH DIMENSION.

BY "A DWELLER IN THREE DIMENSIONS."
(MRS. ELIZABETH BROWN DAVIS, WASHINGTON, D. C.).

We live in space of three dimensions. We call these three dimensions length, breadth, and thickness. For example, a line has length, but no breadth or thickness. A square has length and breadth, but no thickness. A cube has all three—length, breadth, and thickness. All the objects which we touch and use have these three dimensions, no more and no less.

Even when we say that a line has length, but no breadth or thickness, in reality we have to exercise our imagination to picture a line absolutely devoid of breadth or thickness. In practice, if we attempted to make such an object of only one dimension, which we could pick up and handle, the nearest approach to it that we could make would be an extremely fine rod or wire, but the most finely attenuated wire that could possibly be manufactured would evidently have some breadth and some thickness, though they might be extremely minute.

If we attempt to manufacture a surface having two dimensions, length and breadth, but no thickness, we will find it equally impossible. Some of the metals are capable of being rolled into extremely thin sheets, but it would not be true to say that they have no thickness at all. We may speak of the surface of a sheet of paper, but we cannot separate this surface from the paper without taking away some of the thickness with it.

Hence we see that the objects with which we are surrounded on all sides and which we constantly use, all have three dimensions. Our own bodies have three dimensions, and we live in a world of three dimensions. The notion of three dimensions is one of our inherent ideas, bequeathed to us by our earliest ancestors. Hence it is difficult for us to conceive the possibility of a world in which there are either more or less than three dimensions.

It is possible, however, to picture in the imagination a world of two, or even of only one dimension, because to do so, it is only necessary to take away, in imagination, from known objects, a portion of themselves, that is, one or two of their known dimensions, and to picture their appearance as it would be under those conditions.

On the other hand, to picture in the imagination a world of four dimensions, or even one object of four dimensions, requires that we add to three dimensions already known, other parts about which we know nothing whatever. It is obviously much easier to imagine a known object stripped of some of its known parts, but whose remaining parts are also known, than it is to imagine that same known object, with all of its known parts intact, and increased by other parts which are entirely unknown, and about which we have no information to guide us.

Moreover, we have no good reason for supposing that a world of four dimensions does anywhere exist. But the question has often been asked, If there are three dimensions, why are there not four, or five, or even more? Why should the number of dimensions be limited to three? Why should it be limited at all? To this there is clearly no satisfactory answer.

Because a condition, or a state of affairs, has never come within our own experience, does not by any means prove it impossible. There are many things in the world around us to-day, even in daily use, which not many years ago we would have declared impossible. We can readily call to mind several instances of this fact.

Hence, if we are not prepared to admit that a fourth dimension is impossible, we must conclude that it may somewhere, under some circumstances, be a possibility. When we have reached this conclusion, the mind eagerly begins to wonder and question what appearance an object of four dimensions would present, and what would be the conditions of life in a world of four dimensions. Since we have no information to guide us, we must look to the imagination for our only answer, and the imagination is ready to respond, as it always is when called upon, though in this case it has extremely meager data.

The best way to approach the solution of this interesting question, is to picture in the imagination beings of two dimensions, living in a world of two dimensions, and then to imagine the relation of our world of three dimensions to theirs. From this we can reason forward, from the known to the unknown, and by analogy, form some notion of the comparison between our three-dimensional space and a world of four dimensions.

A world of two dimensions would lie in a single plane, having length and breadth, but no thickness. Let us suppose this plane to be horizontal, like the flat top of a table. All the objects in it would be absolutely flat, without any thickness whatever. If such a world of two dimensions were peopled by intelligent beings, their bodies also would have two dimensions, length and breadth, but no thickness. They might have straight sides, like squares or triangles, or they might be curved, but whatever their shape, they would be perfectly flat.

They could glide about the plane in any direction they pleased, as long as they remained in the plane, but they could not move out of their plane. Hence they could not lift themselves up on edge, as we would stand a card on its edge on the table; nor turn themselves over, as we would turn up the face of a card. They could not move one hair's breadth out of their plane, for if they did they would at once be in three dimensions, and we are supposing them to live wholly in two dimensions.

They not only could not *move* out of their plane, but they could *see* only objects lying in their own plane. That is, their eyes would be so constructed that they could see horizontally in every direction in their own plane, but they could see nothing above their plane, and nothing below it.

Instead of imagining their plane a small one resting on the top of a table, we may, if we wish, imagine it a huge plane out in space, reaching out to the most distant stars. They might then be able to see the stars which happened to lie in this extended plane, but no matter how bright the stars not lying in their plane might be, those stars would be invisible to them.

Not only would these creatures be unable to move themselves out of their two-dimensional world into the third dimension, and unable to see any object not lying in their own plane, but their ideas would be equally as limited as their powers of locomotion and of vision. It would be impossible for them to imagine an object having more than two dimensions, and the expression "third dimension" would be as unmeaning to them as the expression "fourth dimension" is to us. For instance, they might understand perfectly all the properties of the square, triangle, and circle, but they would have no conception of a cube, a pyramid, or a sphere, and if any one attempted to describe such objects to them it would be impossible to convey the correct idea to their minds.

Thus we can see how such creatures might live, throughout the entire history of their race, in a world of only two dimensions, seeing and understanding only two dimensions, and yet with three dimensions lying all about them, extending out to infinity above their plane, and to infinity below it.

Now, if there is a fourth dimension, it must encompass the three dimensions with which we are familiar, in very much the same way that three-dimensional space surrounds the plane of two dimensions.

If we should try to explain to the being who knows only two dimensions the meaning of the *third* dimension, we would probably begin by talking to him about one dimension, which, of course, he could easily understand. We would point out to him that if a straight line be drawn in one dimension, and then a second line drawn at right angles to the first, the two lines thus drawn would represent two dimensions. This he would understand perfectly. We would then pass to the next step, and explain to him that, starting from that same right angle, if we construct a third line perpendicular to both of the original lines at their point of intersection, we should then be in space of three dimensions. He would probably be able to follow the reasoning readily, but when he tried to form a picture in his imagination, it would be impossible for him to see how three lines could be perpendicular to one another at one and the same time and at the same point. It would be beyond his utmost power to trace this third line in space.

Practically, this same difficulty confronts us, when we try to pass from the notion of three dimensions to the notion of four dimensions. We know that two lines at right angles to each other lie in a plane of two dimensions. And we know that a third line can be constructed in such a manner that all three lines will be perpendicular to one another in the same point, and that the three directions in which these lines extend will represent the three dimensions of our space. All this is very familiar to us. Now, if we proceed one step further, and construct a line meeting these three lines in their point of intersection, and perpendicular at one and the same time to all three of them, this fourth line will extend in the direction of the fourth dimension. We can follow the reasoning to this point, but when we try to construct the last line, we are in the same position as the being in two dimensions, who could not imagine what direction the third perpendicular would take. When we have found out how to draw four lines, meeting in a point, each of which shall be perpendicular to all the other three, we will have solved the problem of the fourth dimension; or at least we will be very warm, as the children say.

The square of any number, a, is written a^2, and it may be represented graphically by a flat surface bounded by four equal straight lines, whose length is a, and by four right angles. This requires only two dimensions.

The cube of the same number is written a^3, and is represented graphically by a solid figure of three dimensions, bounded by six squares, each equal to a^2. Its angles are formed by three edges meeting in a point, each edge being perpendicular to the other two.

Following the same analogy, the fourth power of the same number is written a^4. This much we know; but its graphic representation we can only imagine, since it could only be formed in four dimensions.

It seems reasonable to suppose that it would be a figure bounded by cubes, since the cube was bounded by squares, and the square by lines; and that its angles would be formed by the meeting of four edges, each perpendicular to the other three.

Let us return for a moment to the consideration of the world of two dimensions, which we have supposed to be a plane resting on some flat surface, as a table, and peopled by flat creatures of two dimensions. It is obvious that in their eyes the edges of objects would constitute the exterior of the objects. We know that if we look down at a card lying on a table, we can see one entire side of the card. But a flat creature, in the same plane with the card, would be able to see only the edges of the card. Even their houses, like everything else, would be flat like the card, and the walls of these houses would be their edges. When their doors were closed, those on the outside would see only the edges or exterior of their card-like houses. And they could not comprehend how we, looking down from our three dimensions, could see the whole interior of these closed houses, just as they would fail to understand our ability to view the entire surface of the card. In some such manner, it might be possible for a creature in the fourth dimension to see the interior of our own houses even when all doors and windows are closed.

If we should purloin an article from one of the two-dimensional closed houses, and remove it entirely from their plane, it would become suddenly invisible to them, and its disappearance would doubtless constitute a great mystery. In the same way, if there were a fourth dimension, it might be possible for some object belonging to us to disappear suddenly and mysteriously into the fourth dimension.

Although the creatures of our hypothetical two-dimensional world would be perfectly flat, they would possess a right side and a left side, just as a person in a photograph has a right and a left side. If we should lift a two-dimensional being from his plane, and replace him in a position that would be from our point of view bottom side up, his right and left sides would be reversed. This may be verified by experimenting with a face card. Hence we may imagine the possibility of any object being lifted from three dimensions into the fourth dimension and replaced in its former position with its right and left sides reversed.

We are told that there are light rays which are invisible to us, solely because our eyes are so constructed as to be unable to perceive them. And we are also told that there are tones so low or so high that we can never hear them, because our ears are not attuned to them. Shakespeare expresses this idea in the Merchant of Venice, when he makes Lorenzo say:

"There's not the smallest orb, which thou beholdest,
But in his motion like an angel sings.
Such harmony is in immortal souls;
But, whilst this muddy vesture of decay
Doth grossly close it in, we cannot hear it."

And hence, whether or not a fourth dimension does really exist, it might be that causes similar to those just mentioned, that is, the limitations of some of our senses, would operate to render us unable to perceive it. But just as we may enjoy in imagination the "music of the spheres," though we cannot hear it, so we may take pleasure in exercising our ingenuity in picturing the different properties of the fourth dimension.

XI.

SOME FOURTH-DIMENSION CURIOSITIES.

BY "CREPUSCULA SUBLUCENT" (G. M. ACKLOM, M.A., NEW YORK CITY).

Before commencing any explanation of what is popularly — i. e., physically — meant by a "fourth dimension," it is necessary to preface that the expression is often used in an entirely different—i. e., a mathematical—sense, which bears no relation to the conception of an actual universe of four dimensions.

Mathematically speaking, the fourth dimension is merely a device of demonstrable utility for the solution of problems in geometry and algebra concerned with more than three independent variables, and is simply a convenient expedient of the same character as $\sqrt{-1}$, a^{-n}, or any other quantities impossible of actual conception, which yet, through the allotting to them of meanings which do not conflict with the laws of real numbers, we are able to use, and find of great service in the extension of mathematical operations. For instance, a given particle of gas would require an expression of four algebraical dimensions if, in addition to its location in space, its density is to be considered. In geometry a line possessing one dimension (length) being moved at right angles to its direction, traces out a surface, of two dimensions; a surface, moved at right angles to the directions of both its dimensions, traces out a solid, of three dimensions; and could we but find a direction at right angles to all of the three directions of these dimensions, a solid moved in this new direction would, we must suppose, trace out a figure of four dimensions. We are not able to visualize such a figure, but it is a simple matter to predicate some of its characteristics. For instance, the projection of a cube may be made on to a plane, or even on to a line; similarly, a "tesseract" (the name given to the fourth-dimension figure traced by the motion of a cube) may be projected on three-dimension space, or even a plane.

Draw the complete diagram of a cube, edge 1 inch; at a distance less than 1 inch away in any direction repeat the diagram; then join all the corresponding points of the two figures, and the result is the plane projection of a tesseract. From it may be observed that, just as a cube is determined by 8 points and 12 lines, and bounded by 6 squares, so the tesseract is determined by 16 points, 32 lines, 24 squares, and bounded by 8 cubes.

Three-dimension webs, or projections of four-dimension figures in space, can with a little more difficulty be made; in fact, have been made in Germany by Dr. V. Schlegel.

Now, in order to gradually attain to the conception of a physical universe of four dimensions, we may consider that an infinite number of lines—i. e., one-dimension figures—laid side by side make up a plane, i. e., a two-dimension figure; and an infinite number of planes laid one on top of the other form a solid, i. e., a three-dimension figure, and therefore, by analogy, an infinite number of three-dimension figures allocated in the requisite direction (if we only knew how to do it) would compose a four-dimension figure.

Or, putting the idea in a slightly different form, a line may be considered as an infinitely thin slice of a surface, a surface as an infinitely thin slice of a solid, and so a solid merely as an infinitely thin slice of an extra-solid (four-dimension figure). Expanding this idea to the whole universe, we see that it follows, as a matter of course, that an infinite number of two-dimension universes is capable of being contained in our space, and, similarly, a universe of four dimensions would of necessity contain an infinite number of universes such as ours.

If we figure to ourselves the conditions of existence in a world of two dimensions, and note the relation such a world would bear to the three-dimension world in which it might lie, we shall get some instructive analogies to the relations which would obtain between our universe and a universe of four dimensions which may be conceived to enfold it.

For instance, the world of an oyster or that of a thin, flat non-burrowing worm would be approximately two dimensions; while, by a *reductio ad ultimam* of these approximate conditions, and by supposing the worm incapable of cognizance or motion up or downward, we may obtain a very fair representation of life in a mathematical plane.

For the sake of economy in words, it will be well to call the two-dimension universe we are examining a plane, the three-dimension universe in which it lies space, and the supposititious enveloping four-dimension universe extra-space; and also to designate their inhabiting organisms by the symbols P, S, and E respectively.

Now, it can be easily seen that for P a line in his plane forms an insuperable barrier, since he is capable of no up or down motion, just as a wall of infinite height would be for S in space.

In the accompanying diagram (Fig. 1), if P wishes

to move from the position *a* to the position *b*, it is obvious that he will have to move round the end of the line *xy*, which, from the nature of the case, he can neither see over nor across; while the object A inside the closed quadrilateral *rs* is as invisible and as inaccessible to him as it would be to S if contained in a closed room in space.

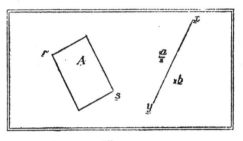

Fig. 1.

Now, it is also perfectly clear that to S, who may be imagined to be occupying a position in space immediately above this plane, the line *xy* is no barrier, should he wish to move from the spot *a* to the spot *b* along the plane, which offers a perfectly free and uninterrupted field for his movement and vision; so that S can, by picking the object A up out of the plane an infinitely small distance into space, and putting it down in the plane again outside of *rs,* render it—as by a miracle—both visible and accessible to P (at *a*). Thus, the whole of P's world lies open and defenseless to the vision and active interference of S. Nothing can be so covered or walled up as to be hidden from him or out of his reach.

In a precisely analogous manner we may imagine that E, from the mysterious recesses of his extra-space,

would be able to act at will on S's world, and to see
everything that S imagined to be hidden. A letter
locked in a safe in a barricaded cellar is as easily seen
and removed by E as the object A is by S. Anything
whatever in space within E's reach may be made to
disappear instantly by the simple process of E moving
it one-billionth of an inch into extra-space, whence he
could return it to space in some different spot—or not
—at his pleasure; and that without any interference
with the integrity of the box, room, or receptacle in
which the object may have been originally contained,
for E does not have to penetrate it, merely to step into

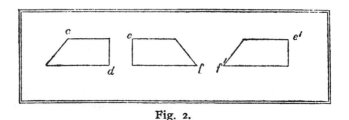

Fig. 2.

extra-space from the outside of the room and back
again into space on the inside of the room. Again,
consider two figures, *cd* and *ef,* in the plane (Diagram
2) which have their sides and angles equal in every
respect. It is clear that P may convince himself of
their identical equality of sides and angles by measurement, but by no possible amount of turning *ef* about
can he make it congruent to *cd;* i. e., capable of being
put in such a position that it can be made to coincide
with *cd* by superposition.

Yet S can do this (to P) impossible thing, by taking
ef up into space, there turning it over, and replacing

it in the plane (as $e'f'$)—a figure bearing the same relation to ef as its own reflection in a mirror.

An exactly analogous process may be performed by E with a solid belonging to space.

For instance, suppose g and h to be two pyramids, irregular but exactly symmetrical to each other and on equal bases, as in Diagram 3. It is obvious that we

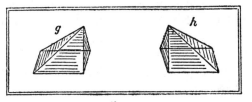

Fig. 3.

may by measurement and calculation establish their equality of cubic content; but by no conceivable turning about of h can we make it fill the same actual space as g (supposing, of course, that g is removed at the time).

But nothing would be easier for E than to take up h into extra-space, there turn it about, and return it to space. Now it will in every way be not only equal to g, but exactly congruent to it and able to fit into the exact portion of space occupied by g. A somewhat similar action is performed in space when we turn a right-hand glove inside out, and so make it exactly congruent to the left-hand one; whereas, previously, it was only perfectly symmetrical to it.

From these considerations we deduce that any body in space which is symmetrical to another can, by being turned about in the fourth dimension, be made identically equal to the other, and symmetrical to its own previous self.

Once more, imagine S to pass completely through
P's plane, and consider how the process will affect P's
consciousness. Of course, only a section of S can exist
in the plane at a given moment (though it is quite possible to conceive of every portion of S being in the plane
at one moment or another during the passing) ; consequently, by no possible means can P become aware that
S is anything but a plane. At the same time, there is
little doubt that S's passage would present some inex-
plicable features to the observation of P.

To begin with, unless S happened to be in the
form of a right prism or cylinder, and struck the plane
with its base exactly parallel to the plane, the section
presented to P's observation would vary continually in
size and contour during the passage.

Even such a regular solid as a sphere would appear
as a circle of gradually increasing and then diminishing
circumference, while if we consider an involved solid,
such as a piece of rope with a few knots, coils, and
hitches in it, we can see that P might well be hard put
to it to comprehend that these alternately separated
and conjoined, irregular, and perpetually varying areas
in his plane were parts of a single whole; let alone the
fact that this whole (could he only know it) is mean-
while in reality not changing its shape at all, merely
its position in space.

The same difficulty will obviously be present to S in
the passage of E or any other fourth dimension body
through space. S can in no possible way become cognizant of more than a section of E at once, and that
section must appear to him as a solid; possibly of
varying size, shape, and character, and possibly also
disappearing and reappearing (owing to extra-spatial
convolutions) as several distinct bodies. In any case,

it will be excessively improbable that S will form any adequate conception of the shape or nature of E as a whole, or even be in a position to recognize E at any future apposition, for his doing so will presuppose that he has encountered exactly the same infinitesimal section of E as before, presented at precisely the same angle in space.

It is not easy at first to conceive of any circumstances under which a section of a body must be a solid, but a glimmer of this possibility may be arrived at thus: Consider the gradual growth of some fixed body—a melon, say—from its birth in the flower to its full development as a large fruit. Every day we have a slightly larger and slightly different shaped solid than we had the day before. Now, taking up a position alongside the melon, think of *time* as a fourth dimension, and visualize in a row the successive shapes which the melon has assumed since its birth. Thus, looking back along the direction of time—as it were—we can mentally become aware of a figure similar to an enlarged elephant tusk, and made up of a vast number of slightly varying and gradually increasing melon-shapes imposed one on the other, and each growing into the one beside it. This figure, observe, does not exist in space, for there has never been more than one melon-shape, as far as our actual senses are concerned, but it may be conceived to have a very real existence in time —our supposititious fourth dimension; and, moreover, it possesses the property of a fourth-dimension figure, for at any given second, i. e., when a section of the figure in time is made, the section appears to us as a solid—a melon. Were we capable of experiencing two widely separated moments in time and their connecting moments (all simultaneously), our melon—and indeed every other growing thing on earth—might appear to us in the manner here conceived of.

Again, in thinking of P's world, we have imagined the plane to be—so far—a rigidly level superficies; but it is quite apparent that, since P is incapable of cognizance of any motion or object outside his own world, this superficies might be curved, in space, without affecting him or his surroundings, or even interfering with the correctness of his scientific observations, since those observations take place exclusively in the superficies; just as the curvature of a sheet of paper will not vitiate the accuracy of a demonstration in plane geometry previously drawn on it. P's world might even be rolled up on itself, so that two places or beings which are an enormous distance (measured in the plane) apart, may be infinitely close to each other when the measurement is made along what S will see as the shortest line, i. e., the distance separating them in space.

Applying this conception to space, we can see that S's world, in a precisely analogous manner, might be curved, twisted, or even involved, in extra space, without S having any possible means of becoming aware of the fact.

This, of course, opens up the possibility that two bodies—say the earth and sun—which are, measured in space, millions of miles apart, may, if the measurement were made along the axis of the fourth dimension, prove to be close together, or actually touching. Some such explanation as this has, as a matter of fact, been invoked occasionally to account for various phenomena, such as the action of various natural forces across vacua of infinite extent, telepathy, and the like.*

---

* If our space were thus curved certain places might be actually much nearer in space of four dimensions than they are in our space, but the difference would in most cases be very slight.—H. P. M.

But just as Euclidean plane geometry fails utterly on an irregularly curved surface where the three angles of a triangle are not equivalent to two right angles, so our solid geometry would prove to be fundamentally incorrect in any portion of space which had such a curvature in extra-space.

Thus we may see that by extending our ideas of the possibilities of existence downward to a world of two dimensions, it is quite possible, if not to obtain actual information, at least to get glimpses of what relations the fourth-dimension world would bear to our universe, supposing it to exist; though it is only fair to say that just as a mathematically plane world would be utterly incapable of apprehension by our three-dimension senses, so a world of four—or more—dimensions, while not impossible of conception, would be equally beyond the reach of our present faculties; so that worlds of two, four, five, or even n dimensions, may coexist with space in the universe we are familiar with, and we all the while be blissfully unconscious of the fact.

XII.

CHARACTERISTICS OF THE FOURTH DIMENSION.

BY "RICHMOND" (LOUIS W. WORRELL,
WASHINGTON, D. C.)

Consider the following figures:

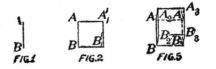

FIG.1   FIG.2   FIG.5

The line AB, possessing but a single dimension, can be moved in a direction not contained within itself, as to the right, so as to generate a surface; for example, square ABB_1A_1. Also the surface ABB_1A_1 possessing but two dimensions, can be moved in a direction not contained within itself, as up from the plane of the paper, so as to generate a solid; for example, cube ABB_1A_1-$A_2B_2B_3A_3$. Can the cube be moved in a direction not contained in its three dimensions so as to generate a new figure whose relation to the cube is analogous to that existing between the cube and its generating square and also analogous to that existing between the square and its generating line? If so, the new figure, called the four-square, contains an additional dimension. This is the fourth dimension.

As the dimension of length is perpendicular to the dimension of width; and as the dimension of height is at right angles to both length and width, the fourth dimension must be perpendicular to the other three.

The above is a full and complete explanation of what scientists call the fourth dimension.

That the possibilities of space are not exnausted with the three dimensions of length, breadth, and thickness has no doubt occurred independently to many minds. However, the present widespread interest in the fourth dimension may be traced directly to Dr. Zöllner, a German astronomer.

Zöllner believed that man is by nature a two-dimensional being; and that he acquires a full comprehension of the third dimension by a purely intellectual process only. The limitations of this article preclude a statement of the process by which he thought that man gains his consciousness of the third dimension. His work shows it to be conceivable that there may be beings who are structurally or intellectually limited to a world of but two dimensions. If structurally so limited, either we cannot imagine them, though we may think about them, or their extent in the third dimension is so extremely small in comparison with their length and breadth that it may be disregarded.

A two-dimensional being living in a plane, as the plane of this paper, might be led to consider the cord,

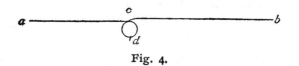

Fig. 4.

ab, with the loop or "knot" *c* lying so close to the paper that he would be unable to see anything unusual about it (Fig. 4).* If asked to untie the "knot" he would move the end *b* entirely around the center of *c*,

* See foot-note, page 30.

whereupon the cord could be pulled straight. If such a being by an intellectual process arrived at a full comprehension of the notion of a third dimension, and attempted to untie the "knot" *c,* he might ask two friends to hold the ends *a* and *b* respectively. Then, after having turned a part *bd* of the cord half way over through the third dimension into this position (Fig. 5):

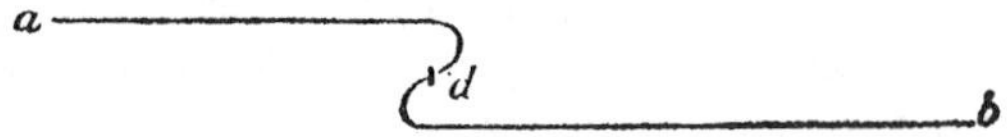

Fig. 5.

he could draw the knotted portion of the cord straight. His companions, seeing him untie the knot thus without moving the end *b,* would be completely mystified by the incomprehensible process. By analogy it seemed plain to Zöllner that some human being, by a purely intellectual process, might arrive at a comprehension of the fourth dimension so complete as to enable him to untie knots in a cord such as this in Fig. 6:

Fig. 6.

without moving its ends, simply by bending some essential part of the knot through the fourth dimension.

This, Zöllner thought, would give a rational explanation of the mystifying rope-untying feats being performed by Slade of England.

Some of the probable characteristics of four-dimensional figures may be determined by analogy. Thus, the characteristics of the four-square are found as follows:

The line has two limiting points, as *A* and *B* in the

figures; the square has four; the cube, eight. For the limiting points, we thus have the series 2, 4, 8. As 16 is evidently the next number in this series, it is probable that the four-square has 16 limiting points.

The line has a single limiting line; the square has 4; the cube, 12. Here the series is 1, 4, 12. The fourth term is found by noting the process by which the square is produced from the line; and the cube, from the square. In producing the square, the line is to be counted twice: as AB in its original position, and as A_1B_1 in its final position. Besides two more lines AA_1 and BB_1 are to be added as being traced out by the points A and B of the line. Similarly, in producing the cube from the square, each of the four lines of the square is to be counted twice: as found in ABB_1A_1 at the beginning, and then as found in the final position $A_2B_2B_3A_3$. Besides four more lines AA_2, BB_2, B_1B_3 and A_1A_3 must be added as being traced respectively by the four points found in the generating square. The rule, therefore, is: Multiply the number of lines in the generating figure by two and add a line for each point in it. The four-square should, therefore, have $2 \times 12 + 8$ lines.

The line has no planes; the square has 1; the cube, 6. Here we have the series 0, 1, 6. By noting how the cube is generated from the square, it is seen that the square is to be counted in two planes: as ABB_1A_1 at the beginning and again as $A_2B_2B_3A_3$ at the end of the generating motion. Besides, each of the *lines* of the square has also traced a plane in the generating process —the plane ABB_2A_2 being generated, for example, by the line AB. The rule founded on the above is: Multiply the number of planes in the generating figure by 2 and add a plane for each line in it. Applying this rule to the square as generated by the line, we find the number of planes to be $2 \times 0 + 1$. The rule thus holding true in the series as far as we can know with certainty, we confidently apply it to the four-square as generated by the cube and find the number of planes in it to be $2 \times 6 + 12$.

Noting that a cube is generated by the motion of a plane, it is thought that the four-square generated by the cube is limited by 8 cubes—each of the 6 planes of the generating cube *itself generating a cube* and there being the two additional cubes formed by the initial and the final positions of the generating cube.

To a being limited either by structure or consciousness to a single dimension, any object, such as the square or the cube, or the four-square, crossing the line on which he lived, would be a wonderful phenomenon. Where a moment before there had been nothing, suddenly a point would appear; and, continuing for a time, it would as suddenly disappear.

Similarly, any object, such as the cube or the four-square, moving along the third dimension and passing through the surface on which a two-dimensional being lived would be to him a marvelous phenomenon. Where a moment before there had been absolutely nothing, suddenly a line would appear. Continuing for a time, it would suddenly and mysteriously disappear.

By analogy, it is reasonable to suppose that whenever a four-dimensional object or being comes within the range of our consciousness, it appears to us as an ordinary solid of three dimensions. Thus, we would perceive the four-square as a cube and nothing else. Likewise, a four-dimensional being moving steadily in the direction of the fourth dimension might suddenly appear at our side within a room destitute of openings. Continuing his motion, the final limiting solid of his body would pass beyond our three-dimensional space into the fourth dimension, and he would disappear as suddenly and as inexplicably as he had appeared.

It has been suggested that possibly many of the small objects each of us loses disappear by rolling out of three-dimensional space into the fourth dimension.

We cannot imagine how beings structurally limited to a single dimension or to two dimensions only, can exist. It is true, we can think about them; but only as being mere abstractions. So far as we know they have no existence. Analogy, based upon the above, says that to four-dimensional beings we are likewise mere abstractions and have no real existence.

If two one-dimensional beings were to meet, they could never pass each other. A being of more dimensions than one might carry one of them through the second dimension around his companion, to the complete mystification of both.

If a two-dimensional being were placed inside a square, he could never get out without breaking through one of the sides. A being of more dimensions than two could, however, lift him through a third dimension and set him down outside of his square without his comprehending in the least by what operation this miraculous result was accomplished.

Similarly, if some of us were locked in an air-tight room we could never get out until an opening were made in one of the six bounding surfaces. But, analogy says that a four-dimensional being might pass us through the fourth dimension and set us on the outside of the room without disturbing any of the bounding walls.

In a line, nothing can be rotated. In a plane, rotation takes place around a point. In three dimensions, rotation takes place around a line. In four dimensions, therefore, rotation takes place around a plane.

If a two-dimensional being were asked to turn *m* into coincidence with *n* in Fig. 7,

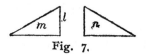

Fig. 7.

he would be unable to do so. A three-dimensional being, however, would simply turn *m* half-way over, through the third dimension, about the side *l*, after which he could easily slide it over *n*.

Similarly, considering our hands, we cannot manipulate them in any way so as to make them coincide. But a four-dimensional being, by rotating one hand half way around about a plane can effect the coincidence easily.

Two one-dimensional beings living on the same line might know themselves to be miles apart, yet they might in the twinkling of an eye be placed face to face if the line were bent into a circle.

Two two-dimensional beings might be miles apart on the surface common to their existence. Yet it is conceivable that a three-dimensional being might bend their surface so as to bring them suddenly together.

Two friends may know themselves to be separated by half the world. Yet it is possible for a four-dimensional being to bend their space of three dimensions so as to bring them suddenly into each other's presence.*

---

* See foot-note, page 142.

Relative to the evidences of a fourth dimension, it may be stated that, as yet, nothing is known which points with any great degree of certainty to its existence.

As stated above, it would be possible to rotate the right hand in the fourth dimension about a plane and thus reverse it and make it coincide with its companion left hand. And because right-and left-handedness is not found in the mass (that is, in mountains, clouds, continents, etc.), but only in the minute (such as is produced in plants and animals and by molecular action), some believe that evidence of this additional dimension must be sought in the region of molecular and cellular activity.

There are two forms of tartaric acid which appear to be identical in every particular except that one turns the plane of polarized light to the right while the other turns it to the left. The right-handed changes into the left-handed without any apparent decomposition and without any apparent manifestation of force. If it can be shown that the change does take place thus, the phenomenon would be proof that the fourth dimension does exist.

In a surface there can be only three points which are equally distant from each other. In space, as we know it, four points and no more can be so arranged. In space of four dimensions five points can be thus placed. Now, in organic chemistry, it has been found that certain substances have the same formulas. For example, there are at least eight possible alcohols which have the formula $C_5H_{12}O$. The only way that chemists have of accounting for these different compounds is by supposing that there is a different grouping of the carbon, or C-atoms in each compound. If now it should become necessary to the explanation of a compound to suppose that five carbon atoms in it are equally distant from each other, this would be evidence of a fourth dimension.

Balance, or symmetry in a line can be produced with reference to a single point only. And, if that point is selected at random, in the line, the symmetry, or balance, can be accomplished by carrying a portion of the longer part of the line through the second dimension around the point and attaching it to the end of the shorter part.

Balance, or symmetry, in a surface, can be produced with reference to a line only. And if that line is selected at random in a figure in the surface, the figure may be put into symmetry only by carrying parts of the longer portions of the figure through the third dimension around the line and attaching them to the shorter portions on the other side.

In a similar way we know that there is three-dimensional symmetry, or balance, with reference to a plane. We see it manifested in the formation of crystals, and in right- and left-handedness, or bi-symmetry of plants and animals. Does this not at least indicate the probable existence of a fourth dimension?

And, finally, Prof. Hinton in developing his mechanics of the fourth dimension came to the conclusion that the mechanics of four-dimensional vortices explain the electric current—a phenomenon hitherto unexplained. He has thus furnished the most direct evidence yet found that space does contain a fourth dimension.

In the seventh book of the "Republic," Plato imagines a group of prisoners chained at the mouth of a cavern. All movement is impossible to them. Their

eyes are constrained to look upon the opposite wall
of the cavern forever. Thus, they never see anything
except their own shadows, together with the shadows
of whatever objects may come in contact with them.
In time, they come to refer all their experiences to their
shadows. And, finally, they identify themselves with
their shadows.

By conceiving a possible state in which man is
limited by his consciousness to less than what he really
is, Plato cleared the way for the notion that the normal
man is likewise limited by *his* consciousness to less than
what *he* really is. This Greater, which Plato strove to
find, is thought by some to involve the fourth dimension.

XIII.

THE FOURTH DIMENSION THE PLAY-GROUND OF MATHEMATICS.

BY "GATH" (ARTHUR R. CRATHORNE, CHAMPAIGN, ILL.).

The fourth dimension has been aptly termed the playground of mathematics. It has certainly called forth much speculation and a great deal of discussion which should not be taken too seriously. To understand the term "fourth dimension," it is necessary to know something of its origin and of the train of thought which led up to it.

If we mark some point on a straight line or on a curve, any other point on it is located by giving one number, its distance from the fixed point. Such a line or curve is called a one-dimensional body, and a given number will locate some point on it. A point may be located on the earth's surface by giving two numbers, the latitude and the longitude. If the streets of a city are numbered, any house may be located by the two numbers which give the house and street. In general, any point on a plane may be located by giving the two distances from two intersecting reference lines. A similar statement may be made for a curved surface. We call such a plane or surface a two-dimensional body, and two numbers will locate a point on it.

The position of an anchored balloon or the bottom of a mine shaft is determined by three numbers: the latitude, longitude, and the vertical distance up or down. A point inside a cube may be located by giving the three distances from the three faces which meet at one corner. Any point in a solid, or more generally any point in space, may be located by three numbers, and conversely any three numbers will locate a point in space. We say then that space is three-dimensional.

Here an inquiring student asked, "Why stop here? Are there points which require four numbers for their representation?" Or the student may be led up to the question in another way. A one-dimensional body, a line or curve, may be the boundary of a two-dimensional body. The boundaries of a three-dimensional body are two-dimensional. "Do three-dimensional bodies bound anything?" Or again, he noticed that if b is the length of the side of a square, then b^2 represents its area, and b^3 the volume of the cube with edge equal to b. "What does b^4 represent? Are there four-dimensional bodies?"

In trying to imagine a four-dimensional thing, the student turned back, and tried to see how three dimensions would appear to a person who knew only two dimensions. He imagined a race of beings endowed with all the faculties of any rational being except that they have but two dimensions and live in a two-dimensional region, say a plane. We might think of these people as the shadows of three-dimensional beings. In their language there are no such words as "up" or "down," "high" or "low." They can see nothing lying outside of the plane in which they live. They can move in any direction in the plane, but have no conception of any movement which will carry them out of the plane. Life in such a region would be under conditions quite different from life in three-dimensional space. A house for such beings may be simply a series of rectangles. A shadow being is just as safe from observation behind a line as a three-dimensional being behind a wall. A bank safe might consist of simply a circle. It would have to be very large, however, for there is no piling up of money in this country. If we imagine a piece of two-dimensional rope, we will see that it is impossible for the shadow beings to tie the two ends together in a knot, even if they had the slightest notion of a knot.

If a schoolboy in shadow land wished to prove that the corresponding angles of the two triangles in Fig. 1 are equal because the corresponding sides are equal, he would perhaps show that each triangle could be moved

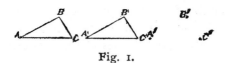

Fig. 1.

over until the vertices occupied the positions $A''B''C''$. He could not place one triangle on the other, for he has no conception of such a thing. If the triangles were as shown in Fig. 2, the schoolboy could not use the

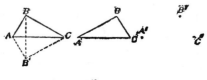

Fig. 2.

sliding method of proof, for no amount of sliding could make the points ABC coincide with $A''B''C''$. He might, however, conceive of the sides AB and BC to be made of some flexible cord, and the point B pushed along the line BB' until the cord again became taut, and then the triangle $AB'C$ could be pushed into the position $A''B''C''$.

In working with this problem, he might have imagined two one-dimensional objects in a one-dimensional region with the fixed points ABC, and $A'B'C'$, respec-

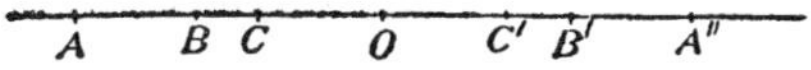

Fig. 3.

tively. These objects may be moved in a straight line, but not out of it. In trying to make the points $A'B'C'$ coincide with ABC, he would find it impossible to do so by sliding along the line, but a rotation about o in the second dimension would bring them together. By analogy he might think that if he could turn his triangle over in the third dimension about AC, he could solve his problem. But he has no conception of such a motion, though he might call his work with the triangle made of flexible cord a revolution in the third dimension.

By a miracle one of these shadow beings becomes endowed with a knowledge of three dimensions. He does marvelous things in the eyes of his neighbors. He can disappear and reappear at will. The strongest prison cannot hold him. If he moves out of the plane in which he has lived he can look down into the houses, even into the insides of his neighbors. If, before returning to shadow land, he should turn himself over, he would be a sort of reflection of his former self to his friends. His heart would be on the right side instead of the left. To his friends he would be left-handed instead of right-handed.

After amusing himself with his two-dimensional people, the student returned to his inquiry as to four dimensions. By analogy he supposed our space of three dimensions to lie in the midst of a space of four dimensions, just as his shadow land lay in the midst of three-dimensional space. He might speak of all people and objects as three-dimensional shadows of four-dimensional things. If now by supernatural means a person becomes endowed with four-dimensional knowledge, he can perform the same kind of antics that his two-dimensional analogue did in shadow land. No prison could hold him. He could take money from a sealed box without making an opening. He could disappear and reappear at will.

In three dimensions we have similar solids which cannot be made to occupy the same space; for example, the right and the left hand. By analogy with the schoolboy's triangle problem, the student conceived of one of such a pair of objects being carried into the fourth dimension turned over and brought back. The two objects can now be made to occupy the same place. Turning a right-hand glove inside out to make it fit the left hand would have the same effect as turning it over in the fourth dimension.*

Since the inhabitants of shadow land have no sense of "up" or "down," they cannot perceive in any way the plane upon which they move but which is present at every point. The imaginative student might then say that the ether which physicists claim to permeate our whole space is but the three-dimensional analogue to the plane of shadow land. So he could go on indefinitely with his analogies, but we must not forget that it is all the product of his imagination, and that there

---

* See Introduction, page 28.

is no more probability of the existence of his four-dimensional beings than of his two-dimensional ones.

While this student was amusing himself with his two- and four-dimensional beings, another student, an investigator in the realm of pure mathematics, had found that the ideas and the language of four dimensions were exceedingly useful. By drawing two perpendicular lines as in Fig. 4, he was able to locate every point in their plane, by giving the distances from each of the two lines. Like the schoolboy who begins his problem, "Let x equal the number of men," the investigator lets x represent the distance of the point from the vertical line and y the distance from the horizontal line. He then, for the time being, concentrates his attention on the letters x and y, just as the schoolboy

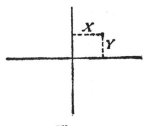

Fig. 4.

manipulates the x in his problem, without at all times keeping in mind that x means men. It is an easy extension by means of three infinite planes to represent any point in space by three numbers, x, y, and z. Again, the investigator, after letting these letters represent the point he is considering, deals only with the letters, and at times pays little attention to what they represent. But there are other things besides points which may be represented by numbers. He may wish to discuss spheres in space. Four numbers are needed to locate a sphere in space, three to locate the center and one to represent the radius. Again, if he wishes to locate a line of given length, he will use three numbers to locate one end of the line; and since the other end can then move on a sphere, he will need two more numbers, or five in all, to represent the line. In any problem he assigns a letter to represent each of his unknown numbers, whether the number helps to give position or not. These letters he uses impartially in his algebraic manipulations. He has called all the points in the plane a "two-dimensional aggregate," for any point is represented by two numbers. The points in space make up a three-dimensional aggregate. The totality of spheres and of straight lines of given length make up four- and five-dimensional aggregates respectively.

These two students are types of the two classes of investigators who have studied the subject of dimensions. The first delights in placing before us those creatures of his imagination, those two and four-dimensional people with their imaginary environment. Just as the dramatist delights in presenting to us a hero who acts, under the conditions laid down in the story, in a manner consistent with his character as presented by the author, so this writer takes pleasure in bringing before our minds his creatures, whose sole characteristics are lack or oversupply of dimensions.

The second investigator is the mathematician who found it a real help in his investigations to use the ideas and language of four or more dimensions. He did not say that a four- or five-dimensional material world existed. He did not believe that our universe was part of an actual four-dimensional space, nor did he ask others to believe it. It was but another example of the mathematician's delight in generalization. In this way he introduced the idea of negative numbers to enrich his language and to give him more power of expression. He never asked us to believe in the existence of a negative number of objects. The chemist is permitted to base his investigations on the atomic theory without knowing or caring much whether such a thing as an atom exists or not. The physicist may talk of the flow of heat in a rod without believing that heat is a substance or that it flows. The mathematician asked to be allowed to extend his notion of space, and to include in it aggregates of more than three dimensions, even if this lead to physical absurdities.

The ideas and phraseology, as exhibited in the writings of investigators in the subject of dimensions, were immediately seized by the romance writers, the prestidigitators, and a certain class of spiritualists. To the first it gave a new method for the disappearance and reappearance of the hero or the villain. As a rule, he returns as a reflection of his former self, having become turned over in the fourth dimension. To the second class it gave a new set of catchwords and phrases for use in sleight-of-hand performances. To the third class, led by Professor Zöllner of Leipsic, the fourth dimension became the abode of the spirit world. For them it solved a great problem, and many are their arguments to prove their contentions. The Bible is brought in to testify, and an extra dimension is read into the meaning of such verses as, "May be able to comprehend with all saints, what is the breadth and length and depth and height." (Ephesians iii, 18.) They boldly stated that physical space lies in a four-dimensional space, just as a line lies in a plane, or a plane lies in three-dimensional space. Just why one should stop at four dimensions is not made clear.

In a brief way we have then shown how the term "fourth dimension" arose. We have shown how the efforts of mankind to tear himself away from the numbers 1, 2, and 3 and to generalize have given rise to two classes of literature, one purely imaginative fiction for the general reader, and one mathematical for the mathematician. From these writings words and phrases have been torn from their settings and used in a way never thought of by their authors, and from this perversion of terms has arisen a discussion which has connected the word "dimensions" with mysticism and the occult.

This, then, is the explanation of the term "fourth dimension." But the persistent reader will perhaps repeat the question, "Is there a fourth dimension?" If by this question he means, "Does a fourth-dimensional world exist physically?" all we can say is that it is highly improbable. Continued thought and discussion on this phase of the question will only result in the state of mind of the Persian poet when he said,

"Myself when young did eagerly frequent
Doctor and saint, and heard great argument
About it and about; but ever more
Came out by the same door where in I went."

If a physical fourth dimension exists a three-dimensional being would never know it, nor would he have any way of finding out. The same statement may be made of two or of five dimensions. As a mental conception, the fourth dimension exists, but the world of our physical experience includes only the three-dimensional.

XIV.

THE TRUE AND FALSE IN THE THEORY OF FOUR DIMENSIONS.

BY "PERGE DECET" (PERCY WILCOX GUMAER, URBANA, ILL.)

Oftentimes, a theory that is advanced in good faith by some distinguished authority falls into disrepute, because it is appropriated by less intelligent persons and is modified or extended to suit some non-related hypothesis of their own. To mutilations of this character the theory of four dimensions has become a sad victim.

The idea originated as a pure mathematical concept, capable of symbolic representation, but quite incapable of being visualized. This may be illustrated by a similar concept found in the use of negative numbers. The individual who subtracts 7 from 3 and gets negative 4 has a mathematical conception of its meaning. He does not, however, infer the actual existence of a negative number of objects. It is easy to conceive that when four trees in a garden are cut down there are four of them missing, yet no person can picture to himself minus four trees, because the mind can visualize only such quantities as result from actual counting. This lack of material existence, however, does not deter anyone from using negative numbers as a short cut in his calculations. In a similar way, the idea of four dimensions may be used in mathematical calculations and without any implication as to the existence of such a space.

Mathematical reasoning has taught us many of the

peculiar properties of this much-discussed space. These properties were appropriated by Zöllner and others as explanations of the phenomena of spiritualism. These persons said that spirits live in a space of four dimensions, and that we human beings who are confined to three dimensions are not sensible of their existence except as they choose to enter our limited space. These statements they have attempted to prove by means of the geometric properties of a four-dimensional space.

In this wise, the unwarranted extension of a mathematical concept has given the lay reader a much perverted idea of the fourth dimension, and it is the purpose of this article briefly to distinguish between the theory as rightfully advanced by mathematicians and the popular conception of the theory after it has been altered to suit the hypothesis of the spiritualists.

In all branches of study or enterprise the mind is greatly aided by concrete representation. Drawings or photographs are indispensable accessories in many branches of industry. No contractor would attempt to erect a building without first securing the drawings for it. So, too, in mathematics, the possibility of drawing a picture of an algebraic equation greatly simplifies its understanding.

Prior to the time of Descartes, the sciences of algebra and of geometry were treated as unrelated subjects. Descartes, however, discovered that algebraic equations of two or of three unknown quantities may be conveniently represented by geometric figures. The method of so representing an algebraic equation can be best illustrated by a simple example. We know from elementary algebra that in an equation of two unknown quantities, such as $y = x^2 - 2x + 2$, we may assign to x any value that we please. Furthermore, by solving the equation we can determine for each value that we assign to *x*, a corresponding value of *y*; for instance, if $x = 1$, we find that $y = 1$; if $x = 2$, $y = 2$; $x = 3$, $y = 5$; $x = 4$, $y = 10$; $x = 5$, $y = 17$, etc. To interpret these results by a diagram, we draw two straight lines meeting at right angles. These lines we call the axes of reference. Along one axis we measure from the intersection distances equal to the

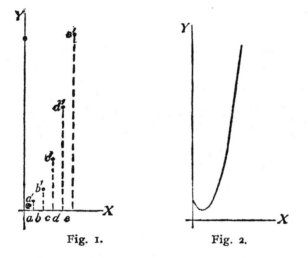

Fig. 1.   Fig. 2.

various values assigned to *x*, as shown by the points *abcde* in Fig. 1. From these points we measure in a direction parallel to the *Y*-axis distances *aa' bb' cc' dd' ee'*, equal respectively to the values that were determined for *y* by the different numbers substituted for *x*. The points *a'b'c'd'e'*, etc., are said to be points on the curve of the equation. It is evident that by assuming the successive values of *x* near enough together we can find an indefinite number of points between those already plotted. Fig. 2 shows the curve of the equation, $y = x^2 - 2x + 2$, as plotted for values of x from 0 to 5.

This concrete representation of an equation may give to the reader but little further information concerning the equation, and the working drawing of any object may be to the layman nothing but a confused mass of lines, yet the drawing conveys to the draughtsman or to the mechanic a very exact conception of the object. So, too, the graph or curve of an equation conveys to the mathematician a concise idea of the properties of the equation.

Sometimes the engineer or the mathematician desires to plot an algebraic equation containing three variables, such as $x + y + z = 10$. Proceeding as before, it is possible to obtain values of z for particular values of x and y. The values of z, however, cannot be represented on the same plane with x and y, for it is necessary to have a third, a z-axis, along which to measure the values of z, and this axis must be perpendicular to the other two at their intersection. Having assumed a z-axis, we proceed to plot the equation of the three variables in the same manner that equations of two variables were plotted. Particular values for x and y are assumed and the equation is solved for z. The values thus obtained for each of the variables are then laid off in the direction of their respective axes.

This concrete representation of equations of two and three variables aided the mind so well in the solution of difficult problems that mathematicians suggested that this interpretation be extended to include equations of four variables such as are sometimes found in problems of electricity and of physics. An equation, such as, $x + y + z + w = 16$, to be plotted, requires a fourth, a W-axis, along which to measure the values of w. Such an axis must be constructed perpendicular to the X, the Y, and the Z axes at their intersection. Here the mathematicians, as the popular saying goes, found themselves up against it, for they could not draw four straight lines mutually perpendicular at a point. This limitation of our space prevented the geometric representation of equations of four variables, but it did not deter further study of the equations.

Men are continually calculating what would happen if conditions were different from what they are. The student of history seeks to determine the effect on history, if Napoleon had won the battle of Waterloo; the physicist calculates the probable amount of heat that would be generated if the earth were suddenly stopped in its orbit; so, too, the mathematician, unable to construct four mutually perpendicular lines, spends valuable time in determining what would happen if it *were* possible to construct his perpendiculars. This leads him to the concept of four-dimensional space.

Here the reader is apt to become confused. The layman, on being told that in a four-dimensional space four straight lines can be constructed mutually perpendicular, immediately seeks to visualize to himself these four perpendiculars. Of course, all such attempts to picture these lines seem futile, and the whole discussion is, forthwith, pronounced a humbug. This, however, is not a fair verdict, because the layman does not usually get the true meaning of the mathematician. It is not meant that these four lines should be actually constructed. That, as far as we are able to know, is impossible. It is perfectly legitimate, however, to calculate what would happen if this were possible, and that is all the mathematician attempts to do.

Physical possibility and mathematical possibility are not always identical. A valid mathematical statement may often be quite incapable of physical interpretation, as will be shown by a reference to Euclid's eleventh axiom. A statement is possible mathematically if it is self-consistent, and if it does not contradict other assumptions in the same discussion. Euclid, the father of geometry, states in his eleventh axiom that through a given point only one straight line can be drawn parallel to another straight line. Proceeding on the assumption that his axiom was true, he built up a system of geometry. In the early part of the 19th century, Lobachevsky, who did not accept Euclid's axiom as true, because it could not be proved, said, let us assume that it is not true. Suppose that through a given point more than one straight line can be drawn parallel to another straight line. He then proceeded by purely mathematical reasoning to build up an entire geometry based on his new axiom. In itself, this geometry is perfectly self-consistent, and it is mathematically possible. Strange as it may seem, we are unable to prove absolutely which system is the true one. Euclidean geometry, however, is simpler, is more convenient, and has been found to hold true even in the most delicate measurements that are possible. Men will continue to use it in their measurements and calculations, because so far as we are able to judge from empirical knowledge, Euclidean geometry is the true one.

So far as our experience goes, all space is three-dimensional, but the statement cannot be proved absolutely. It must be accepted as an axiom. If some Lobachevsky should challenge us for a proof of this axiom, we could give him but little satisfaction. He might then go ahead, and assume that space had four dimensions. He could proceed by deductive reasoning to build an entire geometry based on the assumption that this new axiom were true. He might derive formulæ for the area of triangles, for the volume of solids, or for the direction of a tangent to a curve. This space of four dimensions would be mathematically possible, for all the propositions and deductions concerning it, would be self-consistent and not contradictory within themselves, yet no amount of such reasoning will prove the actual existence of such a space, any more than Lobachevsky proved that any person can really draw through one point two straight lines parallel to a third.

It is possible in dealing with equations of two variables to determine, without plotting them, many of the properties of the curves which they represent. By various manipulations of an algebraic equation, higher mathematics enables us to get the length of any portion of the curve, the direction of a tangent to the curve at any point, or the points of intersection of two curves. The method of studying the properties of a four-dimensional space is very similar to that just described for two and for three dimensions. We know that an equation of four variables represents some sort of a configuration in a space of four dimensions, so that by applying the principles of analytic geometry and calculus to the equation it is possible to determine the properties of the particular figure, solid, or body that the equation represents. It is not at all necessary to be able actually to construct these four-dimensional bodies in order to study their properties. As we determined the properties of curves and surfaces by studying their equations, so we may determine by the same process properties of configurations that are represented by equations of four variables.

Some of the propositions of a four-dimensional geometry are extremely unique and almost incomprehensible. For instance, a hollow flexible sphere in a space of four dimensions could be turned inside out without tearing or stretching.* If any object were capable of moving into a space of four dimensions, it could not be confined by the four walls of a room, and, as soon as it had moved the smallest part of a distance in the unknown direction, it would become invisible. In a space of four dimensions it is possible to revolve an object about a plane, though in three dimensions it is possible to revolve bodies only about straight lines or points.

A study of the strange properties of this hypothetical space, though interesting, is quite beyond the scope of this paper. A geometric proof would require a knowledge of very advanced mathematics, and the wonderful feats that might be accomplished by anyone possessing the secret of a fourth dimension have been well portrayed in several popular articles on the subject.

Is the existence of a four-dimensional space really impossible? is the question most frequently asked. If existence means that the intellectual idea of a thing can be formed, and that this idea shall not lead to contradictions with other well established ideas and with the results of our experience, then it may be said that four-dimensional space does exist. If, on the other hand, existence is taken to mean objective or actual reality, all that we can say about it is that we do not know.

---

* For a mathematical proof of this statement see *Journal of Mathematics*, vol. i, p. 1.

All knowledge proceeds originally from experience, but the amount and the degree of perception possible for our senses is limited. There are many phenomena that are not evident to our senses, and which are known only in an indirect way. We know that there are light waves below the red and above the violet end of the spectrum, which are invisible to the eye. Usually, the non-observation of a phenomena is taken as strong evidence of its non-occurrence. For instance, there was a time when it would have been a reasonable induction to say that all plants and vegetables are motionless, and that animals alone are endowed with the power of locomotion. The perfection of the microscope has, however, shown us that minute plants are as active as minute animals. Hence we cannot always assert that because we do not observe a phenomenon that it does not exist. If we insisted that everything were just as it appeared to be from our observation, we should be in the position of a child who believes that all people have enough to eat, and that all children have nurse-maids. The child reasons from uncontradicted experience, and so do we, usually.

Although we cannot dogmatically deny the existence of a four-dimensional space, even though such a space is inconceivable and impossible for us to imagine, yet we can say with confidence that our universe, as we know it, and every known agency in it, is confined by some unknown law to a space of three dimensions.

XV.

THE ASCENDING SERIES OF DIMENSIONS.

BY "D" (W. S. DAVIDSON, PITTSBURG, PA.).

In setting out to investigate the possibility of a dimension above our present conceptions, we necessarily proceed along the lines of analogy. From comparative investigations in existences of one, two, and three dimensions we will deduce parallel results to enable us to establish formulas by which may be derived an abstract conception of some of the elementary properties of a body in four-dimensional space. To be consistent, we must proceed with the same care with which an astronomer would try to people a remote planet. While considering conditions which make life possible in his own sphere, he would make specific modifications in order to bring it into complete harmony with the new environments.

Although the practical representations of lines and points have appreciable size in all directions, we should not forget that in our discussion these are abstract terms, the latter having only position, and the former, merely a distance between any two positions. In like manner, a surface is imaginary, independent in space, or forming a terminating plane of a body. It is devoid of thickness to such an extent that were an infinite number to be placed one upon the other the aggregate would still have no thickness.

We will consider, first, the limitations in the perceptions of beings in a world of one dimension, that is, existence in an infinite path through space, some portion of which may be represented by the line *AB* (Fig. 1). We will suppose that at various points in this path, separated from each other, creatures *a, b,* and *c* are in progress, *a* representing a point, *b* a creature having length, and *c* a creature similar to *b* but longer. This variety of form is apparent to us because our experience is gained through observing these objects from without the plane of their existence. To the creature *a,* however, *b* is merely a point like itself, and to *b, c* is also a point. This arises from the fact that,

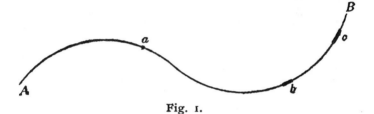

Fig. 1.

having knowledge only of distance and location in their own path, anything requiring a realization of a third quality would be lost to them. Creature *a* observing *b* in the figure would see him as a point, because he sees only the end of *b.* Suppose the bodies, *a, b,* and *c* continued in their relative positions to one another throughout their entire existence, each would then, through his restricted knowledge of the other, be forced to different conclusions regarding the form of life outside his own. Now *a,* conscious of his own existence as a point and observing *b* as a point, would erroneously, though logically, conclude that all life existed in *point* bodies. Creature *b,* upon seeing *a* and *c* as points, and being conscious of his own length, would at once conclude that *he* was especially favored by the Creator above his fellow-beings, apart from the ordinary course of nature.

In Fig. 1, we employed any line, or path in space, but to avoid complications in succeeding diagrams, we shall adopt the straight line. A terminated line may be considered as a path of a point in space, bounded by its initial and final positions. The terminated straight line is the particular case where the point moves from one position to another by the shortest route *AB* (Fig. 2). If we move *AB* through the shortest path to the

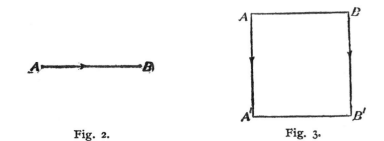

Fig. 2.   Fig. 3.

final location *A'B'*, we obtain a plane figure; and if *AB* be moved in this manner, a total distance equal to its length (Fig. 3), the result will be the definite plane figure known as a square. Hence the square may be called the elementary figure in the two-dimensional world, just as we consider a terminated straight line the elementary figure in the one-dimensional world.

In passing from the lineal to the areal existence, we find that we have greatly multiplied the number of possible varieties in shapes. Thus, our two-dimensional world may have not only creatures represented by points and lines, but also by numerous heterogeneous forms, including many familiar ones, such as the square, rectangle, triangle, and circle. Here, as in the former case (Fig. 1), only by careful demonstration, such as the superimposing of various bodies for purposes of comparison, could these creatures get even an intimation of the endless variety of forms about them, or establish to any definite degree their relationships. An idea of the limited variety of forms that present themselves to the casual observer in such a two-dimensional world, may be gained by us if we cut various-shaped objects from paper and look at them edgewise. A long, narrow strip will appear as

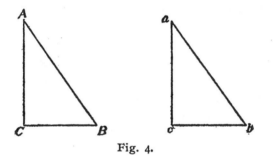

Fig. 4.

a point or a line, according as the spectator views it from the end or from the side; while the square, circle, triangle, and rectangle will appear merely as lines of various lengths. It would be only fair, however, to endow at least a few of these creatures with minds sufficiently mathematical to establish a few simple relationships. Suppose them to be confronted with the problem of proving the entire equality, by the Euclidean method, of the triangles ABC and abc (Fig. 4), when it is granted that the side AB is equal to the side ab, AC is equal to ac, and the angle CAB is equal to the angle cab. The mechanical operation here is within their possibilities.

Let us now place the same figures in the positions shown in Fig. 5. Given the same hypothesis, it might appear at first glace that this case is similar to Fig. 4, but a closer examination shows that Fig. 5 involves the

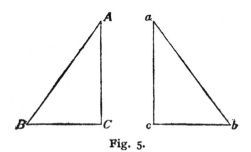

Fig. 5.

simple preliminary operation of reversing one of the triangles before it can be superimposed upon the other. It is evident that this "turning-over" process requires a knowledge of *three* dimensions, and, therefore, to

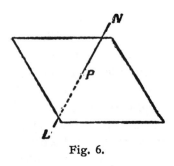

Fig. 6.

creatures with a comprehension of only length and breadth, the possibility of an Euclidean proof would be inconceivable. We will now suppose our two-dimensional world pierced by a line *LN* (Fig. 6), and imagine it to consist of such material that the line may be moved about at will without necessitating its withdrawal from or tearing the plane. It is evident that the only portion of this line which could be detected by these creatures would be the point *P*—a form of creature with which we have supposed them familiar—freely moving about

and apparently limited to the two-dimensional existence, while in reality requiring three-dimensional space
for its accommodation.

We now come to the consideration of objects with
which we are familiar, namely, those in three-dimensional space. All forms of matter manifest to our
senses require space for their accommodation, having
length, breadth, and height. The plane, line, and
point exist in theory only to aid man in the present
crude state of his mental development to build up imperfect images in conformity with forms as he senses
them in the material world. As universal laws are the
media through which nature works, she builds according to conditions and environments and inscrutable
laws of economy. In nature, the straight line and the
plane surface are the exceptions, appearing most frequently among the lower forms of plant and animal
life, but man, ignorant of the finer considerations which
shape the course of nature, and continually prone to
error, must accomplish his results by the simplest, most
direct methods within his comprehension. In doing
this he has adopted as the standard of length a straight
line; the unit of area, a plane figure known as the
square; and the unit of volume, a six-sided figure
called the cube. We have already seen how the plane
may be derived from the straight line, by the same
method we shall construct the elementary figure of
three-dimensional space. Referring to Fig. 3, let us
imagine the square $AA'B'B$ moved at right angles to
its surface, a distance equal to one of its sides. In doing
this we have generated a figure (Fig. 7) which is
three-dimensional.

Suppose that in selecting the straight line AB, from
which our figures have been constructed, we had

chosen one two inches long, then the elementary geometrical figures would have a corresponding mathematical representation, thus: the line $= 2$; the square
$= 2^2$; the cube $= 2^3$. Now since we 'have also such
expressions as 2^4, 2^5, etc., for which we have found no
geometrical solution, the question naturally arises

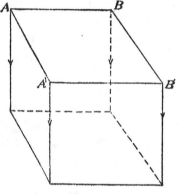

Fig. 7.

whether there does not exist, beyond the limits of man's
present knowledge, a higher order of beings for the
comprehension of whom we, as three-dimensional beings, would require additional perceptive powers.
With our present mental limitations, however, it would
be impossible even to attempt to define an object which
would require four-dimensional space for its accommodation, but by analogy we can deduce a few interesting
facts regarding a figure which would apparently occupy
the same position in the new world that the cube holds
in our own.

We have seen that (1) points form the terminations
of a straight line, (2) straight lines terminate, or
bound, the square, and (3) squares form the bounding

surfaces of the cube. Thus we have established that the elementary figure of each existence is contained within figures having one dimension less than itself. We, therefore, conclude that our four-square figure would be terminated by cubes. In deriving the square from the line, we move along the shortest path from the initial to the final positions, these being separated from each other a distance equal to the line itself. Similarly, the cube was generated by moving the square through space until it occupied a final position at a distance equal to one of its sides from its initial position. In both cases the motion took place in a direction at right angles to *each* and *all* of the boundaries of the generating figure. We, therefore, conclude that our four-square figure might be generated by the displacement of the cube, a distance equal to one of its sides, and in a direction at right angles to *all* of its containing sides. What this direction would be is as foreign to our understanding as a conception of height would prove to creatures in a two-dimensional world.

In the movement of the line to form the square the number of boundaries obtained for the new figure was twice the original, plus two lines generated by the terminating points of the line. In like manner the containing sides of the cube were formed by the first and last positions of the square plus four squares created by the four containing lines. From these considerations it would appear that the four-square would have as its boundaries the initial and final positions of the cube plus six cubes formed by the displacement of the surfaces of the original figure, or a total of eight cubes. Again referring to our square and cube, we see that the number of points or corners in the constructed figure is twice the number of points (or corners) in the generating figure. Thus, the line with two limiting points gives the square four corners; the cube has eight corners and the four-square, on this basis, would have sixteen. The number of lines or edges connecting the corner points is as follows: in the square, twice the original line plus two lines traced by its ends; in the cube, four lines for each position of the square plus four lines described by its four corners. The number of edges of a figure then is seen to be twice the number of lines or edges in the generating figure plus an edge formed by each one of its corners. Therefore, our four-square would give edges as follows: $12 \times 2 + 8 = 32$. To sum up, our four-square would have eight containing cubes, sixteen corners and thirty-two edges; and if our generating cube measured two inches on an edge the content of this new figure would be represented by 2^4.

Curious as the above geometrical deductions may appear, they are surpassed by the dramatic results that would accompany a conception of the fourth dimension. To a creature with a knowledge of mere length and breadth, our physical representation of lines on a plane surface would prove as impassable a barrier as a stone wall unlimited in height would to us. Now, it is evident that we, as three-dimensional beings, may touch all portions of a plane figure (Fig. 3) without disturbing any of the containing lines. If, then, a number of two-dimensional creatures were placed in such an enclosure, imagine their surprise at finding that there existed an order of beings capable of penetrating matter, as they know it, without in any way disturbing it! A parallel case may be imagined in our own existence, if we suppose a being A of the three-dimensional order shut in a hermetically sealed armor-plate vault and suddenly confronted by a being, B, having a knowledge of the fourth dimension.

It might seem possible from these considerations that, with such an advanced state of knowledge, we would be able to extract the pulp from fruit and the kernel from the nut without first removing the outer covering. Likewise, windows for the admission of light, or doors for communication with the outside world, would no longer be necessary, for the fourth dimension would destroy the present effectiveness of the barrier formed by the six sides of a room.

It will be many centuries, if ever, before man can prove the probability of a dimension above the third; but, as we have shown in connection with Fig. 6, we are scarcely justified in denying such an advanced state merely because all matter can apparently be shown to occupy three-dimensional space.

The development of our perceptive senses proceeds very slowly and, according to the theory of evolution, depends upon the extent of the use of existing faculties. We may be justified, therefore, in presuming that we are infinitely nearer to a realization of the four-dimensional existence, if such exists, than we are to the first dawn of reason.

We may consider this ideal state of mental development a possibility if we believe that, in the various stages of his progress, man carries over to each succeeding state a balance of inherent possibilities, which, in the new existence, prove the active influences determining the mental status of the next.

In view of this it might seem possible that that quality of the mind, subconsciousness, is in reality but a subtle force at work evolving greater possibilities in the acquirement of knowledge by the multiplication of the perceptive senses.

XVI.

THE MIND'S EYE AND THE FOURTH DIMENSION.

BY "RAJARAM" (CHARLES JOHNSTON,

NEW YORK, N. Y.)

A straight *line* has length, but neither breadth nor height. It is a figure of *one* dimension or direction.

A flat or plane *surface* has length and breadth, but not height. It is a figure of *two* dimensions or directions.

A *solid* body, like a cube, has length, breadth and height. It is a figure of *three* dimensions or directions.

Line, surface, solid, represent one, two, three dimen-

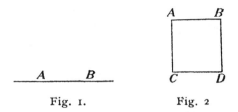

Fig. 1.   Fig. 2

sions. If we could take an additional step, we should have a *fourth* dimension.

In what direction should we look for the fourth dimension? Let us see:

Draw a straight line (Fig. 1). Mark off one inch. This gives a figure of one dimension (length). It is bounded by two points.

On this line as base draw a square (Fig. 2). It has two dimensions (length and breadth). The new dimension is obtained by drawing a line at right angles to the first direction. The square is bounded by four

straight lines; the two-dimensional figure is bounded
by four one-dimensional figures. It has four extreme
points, corners.

From another point of view, the square is formed
by moving the line sideways (at right angles to itself)
for a distance equal to its length.

With the square as base, construct a cube (Fig. 3).
It is a three-dimensional figure. The third direction
is at right angles to the other two, or to any line in the
plane. The cube is bounded by six squares; the three-
dimensional figure is bounded by six two-dimensional
figures. It has twelve bounding lines and eight corners.

The cube is formed by lifting the square upward
from the surface to a height equal to its length or
breadth.

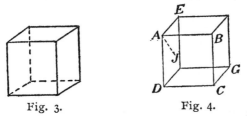

Fig. 3.   Fig. 4.

If we could move the cube in a fourth direction, at
right angles to all its sides, we should form a four-
dimensional figure. By analogy, it would be bounded
by eight three-dimensional figures (cubes), and would
have twenty-four square sides, thirty-two bounding
lines, and sixteen points. (C. H. Hinton calls it a
"tesseract.")

We can represent a three-dimensional figure, like a
cube, on a two-dimensional surface, like paper. It is
just as easy to represent the new four-dimensional
figure on the two-dimensional surface of the paper.

Beginning at the point A (Fig. 4), we draw, first,

AB, a one-dimensional figure. Next, *ABCD*, a two-dimensional figure, *AD* being at right angles to *AB*. Third, the three-dimensional cube, *AG*; its new direction, *AE*, being at right angles to both *AB* and *AD*.

Now, let *AJ* represent a new direction, at right angles to all three directions, *AB, AD* and *AE*. This will be the *fourth* direction or dimension. We can complete the figure as before. This is a true picture of a four-dimensional figure represented on a surface: that is, in space of two dimensions. It is bounded by eight cubes, twenty-four squares, thirty-two lines and sixteen points.

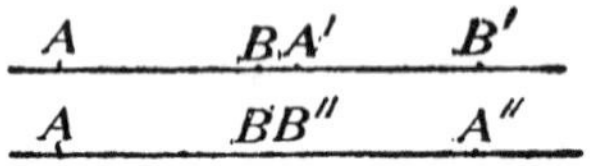

Fig. 5.

Just as the cube was formed by moving the square upward for a distance equal to its length or breadth, so this "four dimensional cube" is formed by moving the cube for an equal distance in a new direction at right angles to all its sides.

In the flat picture (projection) of the cube, the square sides seem to overlap, to occupy the same space. In reality they do not overlap. So, in the flat picture of the four-dimensional figure, the cubes forming its boundaries seem to overlap. But in space of four dimensions they would not overlap.

Let us approach the question in another way. Draw a straight line (Fig. 5). Mark off on it two points, *A* and *B*. A one-dimensional person could push the line along till it reached the position *A'B'*. But he could not rotate it round *B* till it reached the position *B"A"*.

Now, let us take a two-dimensional figure, such as a right-angled triangle, *ABC* (Fig. 6). A two-dimensional man could push the triangle sideways, to the position *A'B'C'* (Fig. 7). He could also rotate it

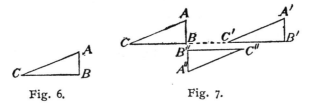

Fig. 6.　　　　　　Fig. 7.

round the point *B* till it took the position *A"B"C"*. But he could not conceive the triangle turned over, so as to take the position *DEF* (Fig. 8). He could not rotate it round a line (*AB*). But for us, three-dimensional folk, it is easy to turn the triangle over—to rotate it round a line—so that it appears reversed, as in a looking-glass.

Now, let us take a three-dimensional figure, a cube

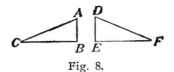

Fig. 8.

(here represented in two dimensions, flat). We can easily think of the cube turned round so as to take the position in Fig 10. We cannot turn it round so as to take the position in Fig. 11, that is, with right and left reversed, as we see ourselves in a looking-glass. We cannot rotate the cube round a surface. A four-dimensional person could, just as we can turn a triangle over, so that right and left are reversed.

We cannot *do* this. But we can easily represent it, either by holding our cube before a looking-glass or by such a diagram as Fig. 12. Here we can think of either the side *ABCD* or the side *EFGH* as being nearest to us, as being the front of the cube. It changes as we look at it.

This right-and-left rotation is characteristically four-dimensional. Something very like it occurs in nature. A beam of polarized light (whose wave-vibrations are all in one plane) is rotated either to the right or the

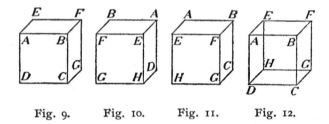

Fig. 9.   Fig. 10.   Fig. 11.   Fig. 12.

left, on passing through certain substances (sugar, starches), just as we might hold a ribbon by the ends and give it a twist to right or left. Dextrose and levulose (forms of sugar found in honey) owe their names to the fact that one rotates a polarized beam to the right (dextra, "right hand"), the other to the left (læva, "left hand"). In chemical constitution, they are exactly the same. Such substances are called "isomeric." It is suggested that their contrasted properties are due to right and left reversal of their atoms, a four-dimensional movement in the minute particles of which they are built up.

Certain snails, exactly alike in all other characters, have a like difference; some are coiled to the right, others to the left. It is remarkable that their juices have a corresponding property of rotating a polarized beam to right or left. This suggests that their external form is an expression of an internal difference, a right or left twist of their atoms, by a four-dimensional force.

The correspondence of the right and left hand, the right and left sides of the face or body, is similar. It could be produced by a four-dimensional twist, just as our Fig. 10 becomes Fig. 11. It is suggested that such a four-dimensional twist runs through living forms; that the life-force is in part four-dimentional

Similarly, it is suggested that electric and magnetic forces are four-dimensional. Let us illustrate: Take a piece of flexible India rubber, shaped like an uncut pencil. You can roll it between your finger and thumb, thus rotating it on its center line (or axis). Now fasten the ends together so as to make a ring. You can turn this ring inside out, rotating it on its axis, which is now a circle instead of a straight line. A smoke-ring has just this motion, turning rapidly inside out. The particles on the outside keep moving to the right, while those on the inside move to the left.

Now, imagine another dimension added to the axis of our smoke-ring. Instead of a circle, it will be a cylinder or tube. The outside surface of the tube will have a continuous movement to the right; the inside, a continuous movement toward the left. It will be a four-dimensional "vortex ring." It is suggested that an electric current going along a wire is such a "four-dimensional vortex-ring." The positive current has a continuous right-hand movement; the negative, an equal and opposite left-hand movement.

Is our mental sight four-dimensional?

Consider the cube in Fig. 12. As we look at it, either face may be taken as the front. Without changing our point of view, we can look at the back and the front equally well; or at the outside and inside of each of the sides. Our line of sight is, therefore, perpendicular to all the sides, as we saw that the fourth dimension must be.

We can do this, because our cube in Fig. 12 is not really solid. A four-dimensional man could do it with a solid cube. And we can do in thought what he could do in fact.

For imagine a solid cube before your mind's eye. You can look direct at the front of it. You can look equally straight at its back or at any side, without either moving your own imagined position or the cube's position. This is four-dimensional.

In the same way you can imagine a locked box, and at the same time imagine the inside of it, without thinking of it as open. You can imagine taking a diamond necklace out of it, while it remains locked. This is four-dimensional robbery, and would be easy to a four-dimensional bank-robber. Our safes would lie open to him for all their locks.

Draw a square on paper. It represents a two-dimensional room. A two-dimensional man could leave it only by going along the surface of the paper to one edge. Put your finger on the paper within the square. It represents the apparition of a three-dimensional being in a two-dimensional room. Raise your finger. The apparition has vanished without approaching the boundaries of the room. Similarly, a four-dimensional being could appear in the center of a three-dimensional room, and disappear as suddenly; just as you can think of yourself in one room and then in another, without having to think of yourself as approaching and going through the doors. This is four-dimensional.

Here is a two-dimensional knot (Fig. 13).* A two-dimensional man could only tie it by rotating half the string in a circle, thus bringing the two ends together. We can tie it by simply folding part of the string over without bringing the ends together. We could also tie such a knot on an endless cord—a circle of string. Similarly, a four-dimensional man could tie one of our three-dimensional knots without bringing the ends of the string together; or he could tie knots on an endless cord—say a ring of leather formed by cutting out the

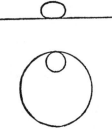

Fig. 13.

center of a disk of leather. The fact that a four-dimensional being could tie such knots, take things from closed boxes, write inside closed box-slates, appear and vanish, suggested to Zöllner of Leipzig that four-dimensional beings do do these things—at séances.

Again, imagine a two-dimensional space, like the surface of water. Take a cone, point downward, and immerse it in the water. First, only a point touches the water. It becomes a tiny circle, which gradually expands, till the whole cone is just immersed. Plunge it deeper, and the cone vanishes from our two-dimensional space—the surface of the water. A two-dimen-

* See foot-note, page 30.

sional man could only conceive a cone in this way: a point, succeeded in time by expanding circles, and finally vanishing.

It is suggested that we in like manner could only perceive a four-dimensional form as a series of three-dimensional forms succeeding each other in time. Thus, the seven ages of man—infant, schoolboy, lover, soldier, to lean and slippered pantaloon—may be thought of as our three-dimensional perception of a four-dimensional form. They may be simultaneous, not successive, like the circles forming the cone. Both vanish at the end—the cone into three-dimensional, the man, perhaps, into four-dimensional space. Thus time, which has only one dimension (length, but not breadth or height), may represent a fourth dimension added to our three-dimensional space.

If our mental vision be four-dimensional, then our mental or spiritual self may be four-dimensional. If séance-wonders are four-dimensional, they may represent the powers of spiritual beings. If time is but the way in which we perceive the fourth dimension, then our spiritual selves, being four-dimensional, may be above time, outside of time—eternal.

Plato may have had this in mind when he compared us to men chained in a cave, watching shadows on the wall. That is, three-dimensional beings limited to two-dimensional perceptions. Did he mean that we are four-dimensional (spiritual) beings limited to three-dimensional (material) perceptions?

Had Paul the fourth dimension in mind when, speaking of spiritual life, he enumerated "the breadth and length and depth and height "(Eph. iii., 18) ; or when he wrote: "I knew a man, whether in the body, or out of the body, I cannot tell, how that he was caught up into paradise, and heard unspeakable words" (II. Cor. xii., 2-3) ? Had John the same thought when he "was in the spirit" and saw "the city foursquare"? Was the body of the resurrection, which appeared in the midst of a closed room, a four-dimensional body? Was the ascension a like disappearance?

These are some of the questions connected with the fourth dimension. This much is certain, that the term comes to us from a firm believer in spiritual life. Henry More, the Platonist, used the phrase "quarta dimensio"—the "fourth dimension"—in his *"Enchiridion Metaphysicum,"* ch. 24, 7, about the year 1671, while Milton was still alive.

Again, solids move in lines, like a bullet; that is, in one dimension. Liquids tend to move in two dimensions, as water spreads over a surface. Gases tend to move in three dimensions, as air fills a bubble. Does ether tend to move in four dimensions? Are its contradictory properties the expression of this?

Are dreams four-dimensional? Is this the reason of their "simultaneous succession"—years in a moment? But our (three-dimensional) space is limited.

XVII.

OTHER DIMENSIONS THAN OURS.

BY "CUBE" (W. T., HOLLAND).

Suppose some men were obliged to creep along inside a long gaspipe, so narrow, that each man would just fit it, and that consequently no two men could pass one another. Then each man would be able to move to and fro in the direction of the pipe, but in no other direction.

In such a case each man would be able to see only the feet of the man in front of him, and if any conversation should be held, it is very probable that it would be about the length they had moved, and not about breadth or height.

If you should look at the pipe from a great distance, you would see it as a black line, and if you were able to see through the wall, you would see little things moving along it.

Suppose a number of men were obliged to creep between two parallel horizontal planes, so near one to the other, that they just fitted between them. They would have more freedom of movement than the men in the pipe, for they would be able to move in different directions. From any given place they would be able to move to some other place by creeping first in one arbitrarily chosen direction and after that in a direction perpendicular to the former.

If you were to stand at a great distance above the planes, you would see only one plane, and if you were able to see through it, you would see little things, seemingly moving on that plane.

In ordinary life our movements are not so restricted as in the above-mentioned states, for we are not only at liberty to move on the surface of the earth, but we can also move in a direction perpendicular to it.

It might be possible to shut up the men within the pipe by means of a small hindrance at each of the two ends. The men between the planes might, in a similar way, be shut up by means of a wire, forming a closed figure, placed between the two planes, say at equal distances from each. Looking at the pipe, you would say that the movement in the pipe is restricted by two points, one at each end of the line, and that the movement on the plane is restricted by some closed figure on the plane.

But in ordinary life, neither a point nor a closed figure (for example, a figure drawn on the surface of the earth) is sufficient to hinder us from moving. We are restricted in our movement only when we are inclosed in a room or some other hollow body.

Now, the mathematician is accustomed to say that a line has one dimension (namely length), that a surface has two dimensions, and that a solid has three dimensions. This is done because a surface may always be compared with a rectangle, which has length and breadth, and a solid may be compared with a rectangular block, which has length, breadth, and height.

For the better understanding of the following, we will suppose that the pipe is really what it seems to be at a distance, namely, a single line, in which creatures are moving, which are not human beings, but which have the form of lines without any thickness.

In the same manner, we will suppose that the creatures between the planes are not human beings, but that they are what they seem to be from a distance, namely, flat figures moving on a plane. The wire, which prevents them from moving at will, is then also a closed figure, drawn on the plane.

In the line, movement is possible only in one direction; therefore, we will call that line one dimensional space, and the creatures therein one-dimensional beings. In the plane, movement is possible in one arbitrarily chosen direction, and also in one perpendicular to that direction; therefore, we will call that plane two-dimensional space, and the creatures in it two-dimensional beings. We ourselves are three-dimensional beings, living in three-dimensional space. In that space we can move in any chosen direction, then in one perpendicular to it, and again in a third direction perpendicular to the first and second. For instance, you may walk along in a street, then move perpendicularly to the street when you enter a house, and after that move perpendicularly to the surface of the earth by rising in an elevator.

Now, the question arises: Is it possible that a fourth direction should exist, which is at the same time perpendicular to the first, second and third direction? That fourth direction we cannot see or draw; we are only able to think and to speak about it. A creature, able to move in the fourth direction, would be a four-dimensional being, and would have at his disposal a four-dimensional space.

A one-dimensional being cannot move in two-dimensional space, but he can think about it; a two-dimensional being cannot move in three-dimensional space, but he can think and speak about it. In the same way, we, three dimensional beings, cannot move in four-dimensional space, but we can, by reasoning, find out what a four-dimensional being would be able to perform, and what things might exist in four-dimensional space. We do that by making a comparison with space of fewer dimensions, as follows:

The one-dimensional space can be supposed to lie in a two-dimensional space; the two-dimensional space to lie in a three-dimensional one. In the same way, three-dimensional space may lie in a four-dimensional one. That is to say, the two-dimensional space surrounds the one-dimensional; the three-dimensional space surrounds the two-dimensional, and thus the four-dimensional space must surround the three-dimensional. A two-dimensional creature would be hindered in its movement by a one-dimensional space lying in its two-dimensional space, if this one-dimensional space were impenetrable; it would be obliged to rest all its life on one side of the line, and it could never come in contact with two-dimensional beings on the other side of the line.

Two two-dimensional creatures on different sides of the line could, perhaps, hear each other, but never see each other.

A three-dimensional creature would be hindered in its movements by a two-dimensional space if this latter were impenetrable; it would be obliged to spend all its life on one side of the plane, and could never come in contact with beings on the other side of the plane.

In the same way, three-dimensional space would, if it were impenetrable from the side of the fourth direction, hinder two four-dimensional beings from coming in contact one with the other.

But as long as our world exists we have never heard

of any hindrance to our movements by some plane. Therefore, we will suppose that one-dimensional space is penetrable for a two-dimensional being; that a two-dimensional space is penetrable for us; and that our three-dimensional space is penetrable for a four-dimensional being. Thus a two-dimensional creature would be able to enter one-dimensional space at any given point, consciously or unconsciously. If for a moment we return to our first idea of human beings in a gaspipe, the two-dimensional creature might (by a groove in the wall) take off the hat of one of the men, and put it a few seconds later on the head of another. Neither the latter nor the owner would have the slightest notion where the hat had come from or had gone to. The former would have lost its hat out of sight immediately; the other would see it appear suddenly. In the same way, a three-dimensional man would be able to take off the hat of a two-dimensional creature, and take it outside the wire-fence, within which the latter is shut up, only by removing it out of the plane and passing it through three-dimensional space. The two-dimensional creature would see his hat disappear, without having any notion of where it had gone to, and a short time after he might see it appear again at a place which he would never be able to reach without breaking the limiting fence. We conclude from this that a four-dimensional being would be able to remove our hat and take it outside the room in which we are, without breaking the walls, or opening a door or a window. We should see the hat disappear without understanding where it had gone to, and would see it reappear after a short time in the street, without seeing where it had come from. The four-dimensional being would have taken it out of our space, and passed it through its own space.

A one-dimensional being is unable to turn so as to make its head occupy the place of its other extremity; but if that same being should be taken to two-dimensional space, it might be turned round there, and then be put in the desired position. In the same way a two-dimensional being would not be able to turn upside down, but a three-dimensional being may do this by taking the two-dimensional being out of its plane, turning it round, and replacing it.

Suppose that same two-dimensional being to have the form of a rectangle with points *ABCD* (Fig. 1), then its fellow-beings would see a very remarkable change in its state. For if before this points *A, B, C,* and *D* succeeded each other the way of the sun they would, after the return of that being, follow each other in a contrary direction. Its fellow-being would have seen it disappear suddenly without having a notion where it had gone to, and would have seen it reappear

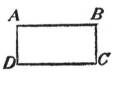

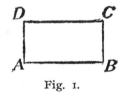

Fig. 1.

suddenly, but now inverted; for what first was on the right will now be on the left.

We gather from this that a four-dimensional being would be able to take a right-hand glove out of our space to his, and to bring it back as a left-hand glove; that a man taken to four-dimensional space might be (but not necessarily must be) transformed into his reflected image, with his heart on the right side of his body instead of on the left side, etc.

In drawing a one-dimensional space we have to draw only lines, each line having two limiting points. In two-dimensional space we can move that line at right angles to its direction, and we can obtain a square. The moving line is in its initial and final position a side of that figure; the two other sides are described by the limiting points of the line. By moving the square perpendicularly to its plane we can obtain a cube. The moving square is in its initial and final position, a limiting face of the cube; the sides of the square describe each another limiting face of the cube; consequently, there are six such-like faces. Each point of the square describes an edge of the cube; the square gives both in its initial and final position four edges; consequently, the cube must have $4 + 4 + 4$, i.-e., 12 edges.

If it were possible to move the cube in a fourth direction, perpendicularly to the three above-mentioned ones, then we should obtain a four-dimensional solid—let us call it an over-cube—of which by comparison we may notice the following properties: The cube in its initial and final position would form a part of the boundary of the solid; each limiting face of the cube would have described a new cube; consequently, the boundary would consist of $2 + 6$, i. e., 8 cubes.

Each of the 12 edges of the cube would describe a square; in both the initial and final position there are 6 squares; so the over-cube would possess $12 + 6 + 6$, i. e., 24 square faces.

Each of the 8 points of the cube would describe an edge; in both the initial and final position there are 12 edges; so there will be a total of $8 + 12 + 12$, i. e., 32 edges.

The one-dimensional line has two final points, the

two-dimensional square has 2 × 2, i. e., 4 points; the cube has 2 × 4, i. e., 8 points; consequently, the over-cube must have 2 × 8, i. e., 16 points.

If a circle in two-dimensional space is passed through a one-dimensional space (which lies in that two-dimensional space), just between two one-dimensional beings, then these beings would be separated. Their distance would gradually increase until the moment that the center of the circle should be in the one-dimensional space; then that distance would be equal to the diameter of the circle. As the circle continues its movement, the beings would be able to approach each other again, and would reach each other at the very moment that the circle should disappear out of their sight. The one-dimensional beings would get the notion of a line, which grows to a certain maximum and then diminishes to zero.

If in the same way a sphere in three-dimensional space were to move through a two-dimensional space then there will be seen on the plane a circle, which gradually increases to a maximum and then diminishes to zero.

We conclude from this that in the fourth dimension there may exist a figure which, by passing through our three-dimensional space, would give us the impression of a sphere, growing larger until it reached a maximum and then gradually diminishing to zero.

For two-dimensional beings that, which we call their surface, is the inner part of them. They are not able to see each other's surface. For one-dimensional beings the line itself forms the inner part; they are not able to see that part of each other. But a two-dimensional being is able to touch the inner part of a one-dimensional being, and a three-dimensional being can touch the inside of a two-dimensional one. In none of these cases the touching creature is seen by the touched one; there can be only a strange feeling.

We conclude from this that a four-dimensional being may be able to touch our inner parts without being seen by us.

XVIII.

THE MEANING OF THE TERM "FOURTH DIMENSION."

BY "GEORGE" (GEORGE GAILEY CHAMBERS, PH.D.).

The phrase "space of four-dimensions" has been used in three distinct connections: in pure mathematics, in various theories put forth to explain certain phenomena in the physical sciences, and, lastly, in attempts to provide a suitable abiding place for the spirits of the dead. It was introduced and developed in mathematics long before it was used in either of the other connections. Moreover, its use in those other connections has been simply a succession of attempts to apply the mathematical concept.

Hence, the aim of this paper is first and chiefly to explain the meaning of the phrase as it is used in mathematics. There it is simply a language device to put certain mathematical facts in a more convenient form or to secure greater generality of expression or for both of these purposes. There is no question raised as to whether such a space actually exists or not. A space of four dimensions arises primarily by generalizing a few of the fundamental facts of ordinary plane and solid geometry. Consequently, an exact explanation cannot be given without first stating those facts on which the generalization is based.

Before taking up that explanation, I will mention some examples of other words whose meaning has been extended in a similar manner. In law, the word person has been extended so as to include a legal corporation. By this device a single statement is sufficient to express any principle of law which applies both to natural persons and to corporations. In double-entry bookkeeping, the accountant charges and credits "bills payable" or "merchandise" just as he charges and credits John Doe. He does this simply as a device which enables him to get a better view of the status of the business. In elementary arithmetic, we use the word "times" in its primary meaning only when the multiplier is an integer, as 3 times $5\frac{1}{2}$. With its original meaning it could not be used to express the related problem of taking a fractional part, as $\frac{3}{4}$ of $5\frac{1}{2}$. The meaning of the word was extended, however, so that we now say $2\frac{3}{4}$ times $5\frac{1}{2}$, or even $\sqrt{2}$ times $5\frac{1}{2}$. We thus secure a generality of expression. This use of the word does not imply at all that anything can happen or be done $2\frac{3}{4}$ times or $\sqrt{2}$ times in reality.

Fig. 1.

The word dimension primarily means measurement. If we think of a straight line (Fig. 1), and of one fixed point on it, O, then the position of every other point on it, P, is fixed by one measurement, if its direction from the fixed point be given. Since one measurement is necessary and sufficient to fix a point, a straight line is called a space of one dimension. This, by the way, is a use of the word space, distinct from its ordinary use. For the same reason, any continuous line is called a space of one dimension. We will distinguish a straight line by calling it a straight one-dimensional space.

In a plane, two measurements are necessary and suffi-

cient to fix a point, P (Fig. 2), with reference to two
perpendicular straight lines OX and OY. Consequently, a plane is called a space of two dimensions.

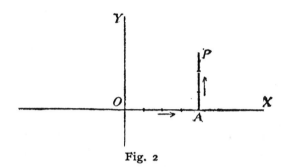

Fig. 2

For the same reason, any continuous surface is also
called a space of two dimensions. We will distinguish
a plane by calling it a straight two-dimensional space.

In like manner, three measurements are necessary

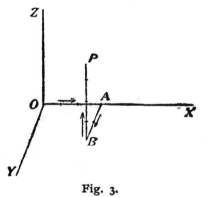

Fig. 3.

and sufficient to fix a point, P (Fig. 3), in ordinary
space with reference to three mutually perpendicular
planes, XOY, YOZ, and ZOX. Hence, ordinary space

is called a space of three dimensions. It should be noticed that this last sentence contains both of the distinct uses of the word space.

We can now state the following definitions which evidently hold in ordinary plane and solid geometry:

A one-dimensional space (a line) is a space such that one measurement is necessary and sufficient to fix a point.

A two-dimensional space (a surface) is a space such that two measurements are necessary and sufficient to fix a point.

A three-dimensional space (such as ordinary space) is a space such that three measurements are necessary and sufficient to fix a point.

We can immediately generalize by adding to this set of definitions the following:

1. A four-dimensional space is a space such that four measurements are necessary and sufficient to fix a point.

Let me here remind the reader that this statement is not intended to define anything that we can conceive mentally in the sense in which we conceive the spaces of fewer dimensions.

In a plane (a straight two-dimensional space) there are an unlimited number of straight lines (straight one-dimensional spaces). In ordinary space (a three-dimensional space) there are an unlimited number of planes (straight two-dimensional spaces). By generalizing we will give to our four-dimensional space the following property:

2. In a four-dimensional space there are an unlimited number of three-dimensional spaces.

The ordinary definition of a plane is as follows: a plane (a straight two-dimensional space) is a surface (a two-dimensional space) such that if any two points in it be joined by a straight line (a straight one-dimensional space), every point in that straight line will lie in the surface. Similarly, we will define a straight three-dimensional space as follows:

3. A straight three-dimensional space is a three-dimensional space such that if any three points in it be joined by a straight two-dimensional space (a plane) every point in that two-dimensional space will lie in the three-dimensional space. Ordinary space is evidently a straight three-dimensional space.

In a plane (a straight two-dimensional space), any straight line (straight one-dimensional space) may be rotated about any point in that line, and even if the amount of rotation be ever so small, the line will occupy an entirely new position, excepting the point about which it was rotated. In ordinary space (a three-dimensional space), any plane (straight two-dimensional space) may be rotated about any straight line (straight one-dimensional space) which lies in that plane, and even if the amount of rotation be ever so small, the plane will occupy an entirely new position, excepting the line about which it was rotated; i. e., any fixed point in ordinary space originally in the plane but not in the axis of rotation will no longer be in that plane. By generalizing we will give to our four-dimensional space the following additional property:

4. In a four-dimensional space, any straight three-dimensional space (such as ordinary space) may be rotated about any two-dimensional space (plane) which lies in that three-dimensional space, and even if the amount of rotation be ever so small, the three-dimensional space will occupy an entirely new position excepting the two-dimensional space (plane) about which it

was rotated; i. e., any fixed point in the four-dimensional space which was originally in the three-dimensional space, but not in the plane of rotation, will no longer be in that three-dimensional space.

From the foregoing definitions and assumptions, the following theorem can be proved:

5. Any four points not all in the same plane determine a straight three-dimensional space.

Proof: Let A, B, C and D be any four points not all in the same plane. Pass a straight three-dimensional space through the points A, B, and C, and rotate it about the plane of A, B, and C. The principle of rotation, 4 above, shows that there will be one and only one position in which the rotating three-dimensional space will contain the point D. Hence the points A, B, C, and D determine a straight three-dimensional space.

From this theorem we have the following corollary:

6. Two straight three-dimensional spaces intersect in a plane.

For if all the points of the intersection do not lie in one plane, let A, B, C, and D be four points of the intersection not all in one plane. Then, by the theorem, there will be just one straight three-dimensional space containing all of them; but by hypothesis they are contained in two such spaces.

Proceeding in this way a geometry of four dimensions can be built up and all the theorems of plane geometry will hold in any plane contained in the four-dimensional space, and likewise all the theorems of solid geometry will hold in any straight three-dimensional space contained in the four-dimensional space. Our ordinary space can always be considered as being one of the three-dimensional spaces contained therein. While the whole structure just described is nothing more than a language device, yet it gives the geometer a means of proving many theorems of plane and solid geometry. In many cases these theorems can be proved much more easily by making use of the geometry of four dimensions than by using the ordinary methods. In fact a number of new theorems in plane and solid geometry have been discovered by means of the geometry of four dimensions. Schubert, in his mathematical essays, gives a very interesting case of that kind.

I wish to refer to one other interesting example before leaving this part of the discussion. It has been proved that in four-dimensional geometry there are six regular structures corresponding to the five regular solids of ordinary geometry. Now, just as a figure in solid geometry can be projected upon a plane, so these regular structures in four-dimensional geometry can be projected upon a three-dimensional space (ordinary space). A few years ago, Dr. Paul R. Heyl, then a graduate student at the University of Pennsylvania, constructed wire models of such projections. These models are now preserved in the mathematical seminar room in the University of Pennsylvania.

The most valuable use of the geometry of four dimensions is distinct from the use mentioned above. To understand it one must have a slight knowledge of analytic geometry, or of the geometrical representation of algebraic equations. Corresponding to any pair of numbers, there is a point in a plane (two-dimensional space); e. g., to the pair of numbers (4, 3) there corresponds the point P (Fig. 2). Corresponding to any set of three numbers there is a point in ordinary space (three-dimensional space); e. g., to the set of numbers (3, 2, 4) there corresponds the point P (Fig. 3). Similarly, from the above definition of four-dimensional space it follows that to any set of four numbers, say (2, 1, 5, 4), there corresponds a point in four-dimensional space.

Also, to any relation in algebra between two variables there corresponds a line in a plane; e. g., to the first-degree equation, $2x + y = 3$, there corresponds a straight line in a plane. To any relation between three variables there corresponds a surface; e. g., to the first-degree equation, $x + 3y + 2z = 1$, there corresponds a plane. Then making use of the language of four-dimensional geometry we can say that corresponding to any relation between four variables there corresponds a three-dimensional space; e. g., to the first-degree equation, $x + y - 2z + 3u = 4$, there corresponds a straight three-dimensional space. This is really nothing but a translation of the algebra into the language of geometry. In a similar manner any algebraic relation can be translated into the language of geometry.

It frequently happens, when a long algebraic discussion is translated into geometric language, that it becomes much more concise, and consequently the mathematician can get a much better view of his discussion as a whole; just as the bookkeeper by using the method of double-entry bookkeeping gets a much better view of the status of affairs in his firm. Moreover, when the bookkeeper has his accounts arrayed by the double-entry method, he is frequently able to discover important facts about his firm's business which would have eluded him if he had used the old single-entry system. Just in the same way, the mathematician has frequently discovered important facts in his algebra by viewing it after translation into the language of geometry. These newly discovered facts can then be translated back into algebraic language and become a valuable addition to his store of knowledge. This is the most important use of four-dimensional geometry.

The value of this will appear to one disposed to look at the practical side, if we consider how these algebraic relations may arise. The problem of a falling body gives rise to a relation between two variables, namely, time and the distance through which the body falls. This gives us an algebraic relation, $S = 16t^2$, from which, by algebraic manipulation, other relations may be derived. These derived relations can then be interpreted in the terms of the original problem of a falling body. In some problems in electricity four variables are related. Such a relation can sometimes be expressed in algebra, deductions made from it, and these deductions interpreted again into the terms of electrical theory. Now, if the mathematician, by making use of the language of geometry, can discover other facts, these facts also can be interpreted into the terms of electrical theory.

Thus far we have treated of the meaning of the term fourth dimension as it is used in mathematics. The same term has been used in attempts to explain certain physical phenomena, such as the phenomena of light. The properties of the space thus assumed by the physicist are exactly the same as the properties assumed or developed by the mathematician. The physicist assumes the existence of a space of four dimensions, takes those properties, combines with them other physical principles, and makes deductions therefrom. He adds nothing to the meaning of the concept of a fourth dimension. Therefore, his theories are outside the scope of this paper.

In the third case, stated in the beginning, that of providing a place for the spirits of the dead, the procedure has been very much the same. Here also no new properties are added to the meaning of the term. The attempts of those interested in this use of it have been directed toward justifying the assumption of its existence. Hence, their considerations also are beyond the present scope.

XIX.

A PUPIL IN GEOMETRY QUIZZES HIS TEACHER ABOUT THE FOURTH DIMENSION.

BEING A REPORT, WITH SOME MODIFICATIONS, OF AN ACTUAL CLASS-ROOM DISCUSSION.

BY "ARCTURUS" (ELMER E. BURNS, JOSEPH MEDILL HIGH SCHOOL, CHICAGO).

Pupil: The newspapers have been printing things lately about the fourth dimension. Will you tell us something about it?

Teacher: I will do my best, but I fear that you will not be able to understand me.

Pupil: I don't understand what the fourth dimension is.

Teacher: State your difficulty as clearly as you can and it may be that I can help you.

Pupil: We have been studying about figures and objects that have length, breadth, and thickness. I don't see how an object can have another dimension.

Teacher: Objects such as you and I can see and handle do not, so far as we know, have a fourth dimension, but there may be other objects that have four dimensions.

Pupil: I don't see how that can be.

Teacher: Well, do you see how there can be an object of only two dimensions? Did you ever see or handle an object that had only length and breadth, but not thickness?

Pupil: No; I never did. Even the thinnest sheet of paper has some thickness.

Teacher: Yet you have no difficulty in dealing with figures, such as triangles and circles, that have no thickness. You even talk about them as though they actually exist.

Pupil: I thought there were such things as circles.

Teacher: You just said that you had never seen or handled anything that had no thickness. Did you ever see or handle a circle?

Pupil: No. Come to think of it, I never did. I see that a circle exists only in my own thought and not in reality.

Teacher: So far as our experience goes, we must admit that is true, but may we not conceive of the possibility of things existing which we cannot see and handle, things beyond the reach of our senses, that have no thickness; in other words, that have only two dimensions?

Pupil: If they are not real to me, I don't see how they can be real at all.

Teacher: Imagine for a moment that your shadow on the wall comes to life. Now, a shadow, as a mere surface, is not real to us. The shadow on the wall is to us a symbol of unreality, of that which has no substance. Can you not now imagine the surface of the wall extended indefinitely and a multitude of such figures as your shadow moving about upon that surface? These shadow figures cannot escape from the surface. They are living in space of two dimensions. If one of them points his finger, he points in some direction in the surface in which he lives; in other words, in the direction of a straight line lying within the plane. Perhaps the earth upon which such creatures live is a circle, and they move about upon the circumference of that circle. Other planets are circles, perhaps moving about a larger circle, as the planets of the solar system move about the sun.

Pupil: Yes, I can imagine all that; but that is a world of two dimensions. I can't imagine a world in which there are four dimensions.

Teacher: Perhaps not, because you have no experience like seeing your shadow to help you. But you may be able to think of the possibility of such a world, and indeed you have taken the first step in that direction.

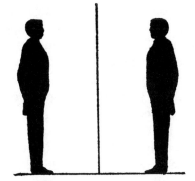

Fig. 1.—The shadowman sees his image in a mirror.

Suppose the shadow man sees his image in a mirror, as in Fig. 1. Suppose, in his vanity, he wishes to appear as he does in the mirror, that is, to take the position of his image. Do you see that, if he goes to the other side of the mirror, remaining in his space of two dimensions; he is either standing on his head or has his back to the mirror? He cannot possibly take the position of his image and remain all the time in the plane. Now, suppose some higher being, who lives in space of three dimensions like yourself, picks up the shadow man and, turning him over, places him in the position of his image—a movement you can represent to yourself by cutting the image out of paper, turning it over and placing it upon the other side of the straight line which represents the mirror. To accomplish this feat, you must take the shadow man out of his own space of two dimensions and move every point of his body in a direction which he himself could not have conceived, because he could not point in any direction which would lead out of the plane in which he lived, and he could not picture to his own mind a direction in which he could not point.

Now, when you look at your image in a mirror, the right and left sides of your body appear to have changed places. That freckle on your right cheek appears on the left cheek of your image. Your image is symmetrical to your body. You have learned in geometry that two symmetrical figures cannot, in general, be made to coincide. You may go behind the mirror as far as the image appeared to be and turn about, yet you cannot take the position of your image, or rather make your body coincide with that position. The freckle is still on your right cheek. Turn about as much as you please in your space of three dimensions, you cannot make your right and left sides exchange places. But suppose there were another direction in which your body might be turned by some higher being, just as you might pick up the shadow man and turn him about in a direction he could not think of, then you might be placed in the exact position of your image.

Pupil: Is that the position I would have if I were in space of four dimensions?

Teacher: Oh, no! That is the position you would have after turning about in space of four dimensions and returning to space of three dimensions, just as the shadow man takes the position of his image after turning about in space of three dimensions and returning to his space of two dimensions. Your whole body would be turned in a direction entirely new to you, a direction in which, so far as you know, you have never yet moved, and a direction in which you cannot point. You can point in all directions in your space of three dimensions, just as the shadow man can point in all directions in his space of two dimensions, but if there is another direction you cannot point in that direction, nor

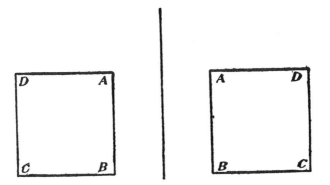

Fig. 2.—The square and its image in a mirror.

can you picture it to your mind, because your mental pictures depend on your experience in space of three dimensions.

Perhaps you can understand it better if we take simple geometrical figures. In Fig. 2 we have a square in the shadow world, and its image in a mirror. It may be placed in the position of its image or the symmetrical position by taking it out of the plane or moving it first in a direction perpendicular to the plane and then turning it over. Tell the shadow man that a line can be drawn at the point B perpendicular to both AB

and *CB* and he will not believe you, because he cannot know of any such direction from his own experience.

In Fig. 3 we have a cube and its image in a mirror. To place the cube in the position of its image, it must first be moved in a direction perpendicular to all its edges. That direction is the fourth dimension. Now you are like the shadow man. When I tell you that

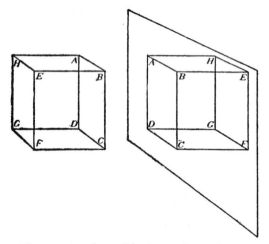

Fig. 3.—A cube and its image in a mirror.

there may be a line at the point *E* perpendicular to the lines *EF, EB,* and *EH,* you do not believe me, because you cannot picture to your mind any such direction.

Pupil: Can you draw that line?

Teacher: That I cannot do. If I draw a line on the blackboard or making any angle with the board, it represents to your mind a line in space, the only space you know, and that is space of three dimensions. You see, the reality of the fourth dimension depends on there being a direction of movement of which we are not conscious. We must admit that, as we know a direction unknown to the shadow man, so some higher being may know a direction unknown to us.

We may unconsciously move in that unknown direction just as the whole world in which the shadow man lives might, unknown to him, be moved in a direction at right angles to the plane.

There is another way in which we may think of the fourth dimension. Just as we can understand how a cube or a sphere appears to the shadow man, as it passes

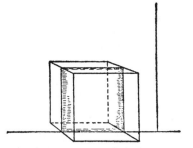

Fig. 4.—A cube passing through the shadow world. To the shadow man it appears as a square.

through the surface in which he lives, so we can understand how certain bodies of four dimensions would appear if they were to pass through our space.

Let us think of a cube passing through the shadow world, as in Fig. 4. The shadow man can see only that part of the cube which lies within the surface in which he lives. If the cube is passing through with four edges perpendicular to that surface, the shadow man sees a square. As the cube passes through, the substance of which that square is composed changes. We can see this if we suppose the surface of the cube to be shaded, say from yellow through orange to red. Now, the colors of the lines which the shadow man sees bounding the square will change as the cube moves through his space. If he is a highly intelligent shadow man he may think of the possibility of a third dimension and try to imagine it just as you are trying to imagine the fourth dimension. He has seen the changing square as the cube moved through his space, and, since he cannot picture to his mind the third dimension, he can only represent this strange figure by a series of squares. Each square is a section of the cube. He would know that in order to change through the entire series of squares the strange figure must move in the third dimension, a distance equal to a side of the square.

We may think of a figure of four dimensions which bears the same relation to the cube that the cube bears to the square. A cube is generated by a square moving in a direction perpendicular to its sides, that is, in the third dimension. So this new figure is generated by a cube moving in a direction perpendicular to all its edges, that is, in the fourth dimension. As a cube moving through space of two dimensions appears as a continuously changing square, so this new figure in passing through our space would appear as a continuously changing cube. As the shadow man represents the cube to himself by a series of squares, each square being a section of the cube, so we may represent this new figure of four dimensions by a series of cubes, each cube being a section of the figure. To pass through our space, this curious figure of four dimensions must move a distance equal to one of the edges of the cube.

We might reason about other figures in the same way. We might even think of a being whose form bears the same relation to the human form that the human form bears to its shadow.

Pupil: I begin to see how we can think of a fourth dimension, but how can you prove that it is real?

Teacher: I cannot prove to you that it is real, since I have never yet seen a body disappear from our space and return to it after turning about in space of four dimensions; nor have I seen a four-dimensional being move through our space. If either of these things were to happen or could be proved to have happened, we should know to a certainty that there is a fourth dimension.

Pupil: Could anyone draw a picture of a body that has four dimensions?

Teacher: If such a picture were drawn it would have three dimensions, just as you may draw a picture of a cube and your picture has two dimensions. If your picture is drawn according to the laws of perspective it represents to your mind a cube, as the cube appears to you. Now, if a picture, having three dimensions and representing a body of four dimensions, were drawn, and if this picture accorded with the laws of perspective in space of four dimensions, still it would not represent to your mind a figure of four dimensions. You would probably mistake it for a model. You would see only the three-dimensional figure. It would require a being conscious of movement in the fourth dimension to interpret the picture.

XX.

POSSIBLE MOVEMENTS AND FORMS IN A SYSTEM OF FOUR DIMENSIONS.

BY "DER CHEMIKER" (J. CLYDE HOSTETTER, LEWISBURG, PA.).

Geometry tells us that a point has no dimension; that it possesses merely position in space. If, however, we move a point continuously in space it will generate a line (Fig. 1), which is said to possess one dimension

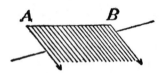

Fig. 1.—Moving point, *P*, through space generates a line which has *one* dimension—length.

Fig. 2.—Moving line, *A B*, generates a surface having *two* dimensions—length and breadth.

—length. Now, let us move the line thus made through space. It generates a surface (Fig. 2), and we notice that our surface possesses the one dimension of the line and also a second dimension—breadth. From a line possessing one dimension we have generated a surface with two dimensions. Now, if we move our surface through space it will generate a solid (Fig. 3). This possesses the length and breadth of the surface and, in addition to these, a third dimension—thickness. From a point, then, we have generated a line with one dimension; from a line we have generated a surface with two dimensions, and from a surface with two dimensions we have generated a solid with three dimensions. We have generated each of these in turn from a form possessing one less dimension by motion through a new dimension. Reasoning from this we conclude that if we could move our solid through a new dimension a figure would be generated which possessed not only the length, breadth, and thickness of the solid, but, in addition to these, still another

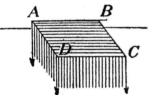

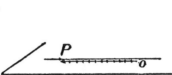

Fig. 3.—Moving surface, *A B C D*, generates a solid which has *three* dimensions—length, breadth, and thickness.

Fig. 4.—To determine point *P* on a line we measure from zero to the point, obtaining *one* number.

dimension. Such a figure would possess four dimensions, and the existence of such a figure would require the existence of a fourth dimension. It is by reasoning of this kind that the idea of a fourth dimension has been developed.

Now, let us take a line and see why the term one-dimensional is applied to it. On a line, the position of a point and, therefore, the point itself, is determined when its distance from an arbitrarily chosen point on the line, the zero point, is known. We find this distance by measuring, in terms of the unit of length, from zero to the point *P* (Fig. 4), in just the same

manner as we measure temperatures on a thermometer
scale, the zero point of which has been arbitrarily
fixed.   One number, then, determines the position
of our point.   Now, a line may be considered as consisting of an infinite number of points.   So any point
of this point-aggregate is determined by one number,
and, in general, a one-dimensional system requires
one number for its determination.   How is the point
determined in a two-dimensional system of points,
such as the plane?   In determining the point on a line
we arbitrarily set a zero.   Here we must also have a
zero for our measurements.   We make this zero the

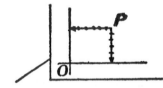

Fig. 5.—To locate *P* on a plane
we measure from *P* to each
of two axes at right angles,
thus obtaining *two* num-
bers.

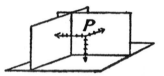

Fig. 6.—To locate point *P* in
space, we measure from *P*
to each of three axial planes
at right angles, thus ob-
taining *three* numbers.

point at which two lines intersect each other at right
angles.   Such lines may be considered the axes of
length and breadth.   Now we measure the distance
from *P* (Fig. 5) to each axis, and having these two
distances our point can be determined.   This is the
same system that is used in locating positions on the
surface of the earth, when we refer distances to the
parallels of latitude and meridians of longitude.   A
point in the two-dimensional plane requires, then, two
numbers for its determination, and, in general, for a
two-dimensional system two numbers are necessary

and sufficient for its determination. The idea of axes is also used in determining position in a solid. A reference to Fig. 6 will make this clear. Here the three distances from point P to each of three planes intersecting at right angles are necessary to determine the position of P. So we find necessary the relation of three distances to determine position in a three-dimensional system. And, to generalize, an "n"-dimensional system of points is such that "n" numbers are necessary and sufficient to determine an individual point amid all the points of the aggregate. Thus, in a fourth-dimensional system four numbers are necessary, and in a fifth-dimensional system five numbers, and so on.

Let us now study the possibilities of motion in the different systems. In a one-dimensional system there is but one possible direction for movement. In a two-dimensional system there is the possibility of mover it in two directions. On a line, then, motion is possible in but one direction; in a plane, motion is possible in two directions. In a two-dimensional system all movements are either parallel to the two axes, or are combinations of movement in these two directions. Similarly, in a three-dimensional system, there is possible motion in three directions, and all movements in a three-dimensional system are either parallel to the three axes of length, breadth, and thickness, or are combinations of movement in these three directions. If, then, we extend the argument, we see that in a fourth-dimensional system, movement would be possible in one or all of four directions.

How many dimensions does the world in which we live possess? We have seen that a solid possesses three dimensions. Further, according to geometry, a solid is a limited portion of space. If we expand our three-dimensional solid indefinitely, it would consequently fill the space. We are accustomed to consider space, therefore, as three-dimensional, and our world is likewise a world of three dimensions. So reasoning as above, every point in space can be reached by motion in three directions.

However, there are some who argue as follows: Motion in one direction will not take us to every point in a two-dimensional system; likewise, motion in two directions will not take us to every point in a three-dimensional system. So, they assert that motion in three directions will not enable us to reach all points in space as it really is. We know that motion in three directions will take us to every point in a three-dimensional system. Then, if motion in three directions will not take us to all points in space, we must assume motion in a fourth dimension, and so a fourth-dimensional space.

What, then, is this fourth dimension, and is there any evidence for its existence? Before we attempt to answer let us see clearly the difficulties encountered in dicussing the fourth dimension. To beings living in a one-dimensional world the idea of breadth has no significance. To beings living in a two-dimensional world the idea of thickness would have no significance. They can move in but two directions and their world is consequently limited to the dimensions of length and breadth. Terms which are easily comprehended by us, who live in a world of three dimensions, would possess absolutely no significance to the two-dimensional beings. Similar to this, then, is the difficulty of describing the fourth dimension. If a fourth-dimensional being were to describe this dimension his description would contain terms having no meaning to us. And when we attempt to describe this dimension we find our vocabulary, developed from our three-dimensional experience, too limited. The best we can do is to discuss the possibilities of a world possessing four dimensions. We can determine some of these possibilities by analogies from our three-dimensional experiences.

The first analogy depends on the properties of configuration. In a two-dimensional system we can place three points at equal distances from one another. Taking a plane as our two-dimensional system, and con-

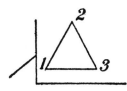

Fig. 7.—In a plane which has *two* dimensions three points can be equidistant, but not four points.

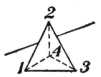

Fig. 8.—In three-dimensional space *four* points can be equidistant, but no more.

necting the three equidistant points, we have an equilateral triangle (Fig. 7). Try as we will, however, we cannot place four points in a plane equidistant from one another. If we add another dimension to our system the placing of four points equidistant from one another can be accomplished. Taking three of the points arranged in the form of an equilateral triangle as a base, we place the fourth point in the third dimension above the others. We can place this at the same distance from the points in the plane, as these are from each other. Connecting our points by lines we have a tetrahedron, the vertices of which are equidistant (Fig. 8). The placing of five points equidistant from one another is impossible as long as we have but three dimensions, but it would be possible if we could use a fourth dimension.

Let us illustrate this space arrangement. In chemistry, the molecules of a compound are said to consist of the atoms of the elements contained in the compound. These atoms are supposed to be at certain distances from one another, and to be held in their relative positions by certain forces. Formerly, all the atoms in a molecule were conceived to lie in one and the same plane. Now, however, the atoms are given a definite space arrangement. In order to account for certain facts, it has been necessary to assume in some molecules that four atoms are equidistant from one another. We picture them, therefore, as being situated at the vertices of a regular tetrahedron. If it were necessary to assume the equidistance of five atoms in the molecule, this would be evidence for the existence of a fourth dimension, as only in a four-dimensional system would this be possible.

Another analogy depends on the properties of rotation. In a plane, rotation takes place about a point; as may be illustrated by the drawing of a circle by means of a compass, in which the end of one leg of the compass is the point about which rotation takes place. It is impossible to have rotation about a point in a one-dimensional system, as a line. In a three-dimensional system, rotation may take place about a line, as, for instance, the rotation of the earth about its axis. In a world possessing four dimensions, however, we see by analogy that rotation would also be possible about a plane.

Let us see if this conclusion is justified. The process of rotation is closely connected with that of superposition, so the latter must be discussed to some extent. The congruence, or, roughly speaking, the equality of two geometric forms is determined by superimposing one upon the other, and then seeing if the two forms

Fig. 9.—Superposition of lines requires *two* dimensions.

coincide in every part. In a one-dimensional system we cannot superimpose one line upon the other;* the best we can do is to place the lines so that they meet. The only way in which superposition of lines can be secured is by moving one of the lines through a second dimension and then placing it upon the other (Fig. 9).

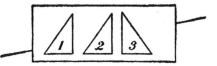

Fig. 10.—The equality of 1 and 2 can be shown by displacement. Keeping 2 and 3 in the plane they cannot be made to coincide by any movement. Rotation of 3 through a *third* dimension makes coincidence possible.

It takes, then, a two-dimensional system to give us superposition of one dimension. Now, take the case of two equal triangles on a plane (Fig. 10). We can determine the congruence of 1 and 2 by displacement; that is, we move one of the triangles and then see if

---

* From what follows it appears that the author means superposition of A′ upon A and B′ upon B.—H. P. M.

the second can be made to occupy exactly the space formerly occupied by the first. But how about triangles 2 and 3? We see that here we cannot use the process of displacement. We can measure the angles and sides and determine their equality, but we cannot superimpose one upon the other as long as they remain in the plane. It required two dimensions for the superposition of lines having but one dimension. Our triangles have two dimensions, and we at once conclude that superposition requires a third dimension. So we

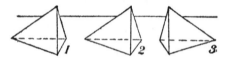

Fig. 11.—The equality of pyramids 1 and 2 can be shown by displacement. 2 and 3, symmetrical pyramids, cannot, in three-dimensional space, be made to coincide. By rotation of one of them through a *fourth* dimension, coincidence would be possible.

rotate one of them about an edge through a third dimension until it again reaches the plane, and they can now be superimposed. Rotation about a line and superposition of two-dimensional figures require thus the aid of a third dimension. In this passage through the third dimension, however, the angles of the triangle were reversed, that is, the anterior and posterior angles are interchanged, and, in fact, it is due to this that the superposition is possible.

Let us extend this idea to the superposition of one solid upon another. For pyramids 1 and 2 we can use the process of displacement (Fig. 11). How can we superimpose 3 upon 2? Such pyramids are symmetrical. All lengths and angles of one have their exact duplicate in the other, yet the two cannot be made to coincide, that is, be fitted the one into the other so that they shall both stand as one pyramid. They correspond exactly to our left and right hands. Our hands cannot be made to coincide in our three-dimensional space. The reflected image of the right hand, however, could be made to coincide with the left hand; they are alike one another, but on opposite sides of a plane. Just so are pyramids 2 and 3. We cannot, in our three-dimensional space cause symmetrical pyramids to coincide. It requires rotation about a plane to give us congruence. This is impossible now, but if it were possible to hold one of the surfaces of either pyramid and rotate the pyramid through a fourth dimension back into our three-dimensional world it could be accomplished. This is the fourth-dimensional analogue of the superposition of the two triangles above described. In this rotation the interior surfaces would be converted into exterior surfaces,* and it is due to this conversion that coincidence is now possible. This interchange of exterior and interior surfaces may be illustrated by turning a right glove inside out to form a left glove.

Now, to take another illustration from chemistry, there are two varieties of tartaric acid which crystallize in forms bearing the relation of object to mirror-image. Such crystals are illustrated in Fig. 12. Apparently these two varieties change the one into the other without chemical resolution and reconstitution. If it could be shown that such does take place, then this would be proof of a fourth dimension, because only in a fourth-dimensional space can a right-handed shape become a left-handed shape by simple movement.

---

* This is not true. See Introduction, page 28.—H. P. M.

These, then, are the most obvious of the possibilities of a fourth dimension.

Is there a real fourth-dimensional world? It is highly improbable. If there were such a world would it be inhabited by beings who could act upon us three-dimensional beings, as the Spiritualists assert? We reasoned the possibilities of a fourth-dimensional world by analogy—we must reason this question in the same way. If there is a fourth-dimensional world containing beings that can act upon and influence us, who are but three-dimensional, then, by analogy,

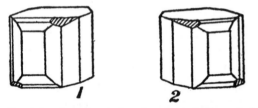

Fig. 12.—Crystals of a tartrate bearing the relation of object and image. If 1 changed into 2 without chemical resolution and reconstitution it would be *proof* of a *fourth* dimension.

we would expect the existence of a world of two-dimensional beings upon whom we could consciously act. We do not know of such a world. Also, we would expect a fifth-dimensional world with beings who could influence the beings of a fourth-dimensional world, and so on. Perhaps there is a two-dimensional world that we cannot influence. Then, the other worlds should be independent also, and if the fourth-dimensional beings can still influence us, then the fourth-dimensional world would be an exception in the great plan of creation. The existence of such a world with beings that can influence us is, therefore, highly improbable.

In conclusion, let us summarize what we have learned concerning the fourth dimension.

In a system of four dimensions:

1. It would be possible to generate a body possessing four dimensions by moving a solid through the fourth dimension, just as a solid is generated in a three-dimensional system from a surface with two dimensions.

2. It would be possible to move in four directions, whereas, now we can move in but three.

3. It would be possible to place five points equidistant from one another, whereas, now four is the maximum number.

4. Rotation would be possible about a plane, whereas, now it is possible only about points and lines.

5. Coincidence of symmetrical solids would be possible.

6. If there is such, it is highly improbable that it is inhabited by beings that can act upon us who are three-dimensional.

Grateful acknowledgment is here given by the writer to Hermann Schubert and C. H. Hinton, whose papers have been freely used in the preparation of this essav

XXI.

THE FAIRYLAND OF THE FOURTH DIMENSION.

BY "A. CLEMENTUS" (A. C. SILVERMAN, SYRACUSE, N. Y.).

Everybody has observed the difficulty that a little child has in realizing that it must step over a comparatively high object on the floor. It has no notion of falling. It is delighted and astonished as, from its eminence on the table, it watches you bend down and disappear and then rise up again and cry "peek-a-boo." This inability on the part of babies to comprehend a third dimension is well known. Now, very serious and bespectacled geometricians tell us that perhaps we, too, are but babies in a space of a fourth dimension, and that we, too, might be no less astonished if beings from that world chose to play peek-a-boo with us.

In order to get some notion of the fourth dimension, let us, first of all, get an idea of the meaning of dimension. The dictionary gives it as extension in space. Every material body, such as a tree, a horse, a sheet of paper, is known as a (physical) solid, and the limited portion of space it occupies is known as a (geometrical) solid, because it extends in three directions; and we speak of every object as having three dimensions—length, breadth, and thickness. Yet, although a very thin sheet of paper is a solid, we can think of its surface only; and, although a tree is a solid, we can think of its height only, without any reference to its diameter.

This is true, for we do have the linear measure with its inch, foot, yard, and we have the square measure with its square inch, square foot, and square yard. Indeed, we may get an idea of the cube by drawing, first, a straight line; then, another straight line perpendicular to the first at its extremity, forming a square; and then a third line perpendicular to the other two at the same extremity, forming a box or cube, the volume of which is expressed by the cubic measure, the cubic inch, cubic foot, and so forth. The same idea can be gotten from the following definitions in geometry: a point has position but no magnitude; if a point moves it generates or traces a line and that has length only; if a line moves, not along itself, it generates a surface, which has length and width; if a surface moves, not along itself, it generates a solid, which has thickness, besides the other two dimensions.

But, having the solid, our experience does not permit us to go any further. However we move the solid, we still generate a solid and nothing else. Nevertheless, let us be bold and imagine that we move the solid into a space that it did not previously occupy and that we make it take an added dimension that it did not previously have. We now get an object of four dimensions.

It may be difficult for us to form a conception of a world of more than three dimensions. Yet it is no more difficult than to imagine a world confined to only two dimensions, or than, for beings of such a world, to form a conception of our space.

For simplicity, let the two-dimensional world be a plane, though equally well it might be the surface of a sphere. We may picture to ourselves the mode of life of the inhabitants of this flat land. They could move in any direction along the plane, but they could not move perpendicularly to it, and would have no consciousness that such a motion was possible. They would not be able to turn their heads up or down. Things about them could be pulled or pushed in any direction, but they could not be lifted up. People and things could pass around each other, but they could not step over anything. Their plane geometry, however, would be exactly like ours.

In this supposed land, let us draw two straight lines perpendicular to one another, that is, two straight lines intersecting at right angles at A. The drawing (Fig. 1) would be as perfectly conceivable to our plane

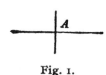

Fig. 1.

beings as it is to us. But suppose we asked them to draw a third line perpendicular to the other two lines at the same point of intersection A. That would seem absurd and impossible to them, just as it would be to us if we had to draw the required third line on the paper. But with this condition removed, we can leave the plane surface of the paper and draw the third line through the paper perpendicular to the surface at A, just as we might stick a pin at A vertical to this page.

So, too, with us, when we have a cube after drawing three mutually perpendicular lines, and are required to draw a fourth line passing through the same point, perpendicular to all of the three lines already there. In our space the problem is absurd and impossible. Our conceptions do not admit of more than three dimensions. But for a being that could conceive of a fourth dimension the problem would be easy. He would simply draw the line through that space.

Our conscious life is in three dimensions, and naturally the idea occurs whether there may not be a fourth dimension. No inhabitant of flatland could realize what life in a world of three dimensions would mean. Yet, if he were intellectual, he might be able to extend the analytical geometry that applied to his world, so as to obtain results true for a world in three dimensions, a world that would be unknown and inconceivable to him. Similarly, we cannot realize what life in four dimensions is like, though we can use our analytical geometry to obtain results true of that world or even of worlds of higher dimensions. Moreover, the analogy of our position to the inhabitant of flatland enables us to form some idea of how the inhabitants of space of four dimensions would regard us.

If we placed a dweller of flatland inside a circle, or inside a rectangle drawn on his plane, he would be as truly imprisoned as we are in a closed prison cell. He would go all around, and, finding every inch of it closed, he would simply despair of getting out, unless he could break through it. On account of his limited conceptions, he could not possibly understand how we might step over the boundary. He could form no notion of the trick. But we should simply step over the line and reappear on the other side. So, if we, confined within the six surfaces of a dungeon, a being able to move in the fourth dimension, he would step outside of the cell without breaking any part of the walls, ceiling or floor. He would do it as easily as we could pass over the circle drawn on a plane without touching it—so wonderful to our friend in flatland. Our new being, the fourth-dimensional one, would simply disappear from our view like a spirit and then reappear again outside the prison. He would only have to pass through the fourth dimension. Of course, no such a case has as yet been reported.

Let us continue our analogy further. We know that the cross-section of a line is a point; that of a surface, a line; and that of a cube, a surface. Hence, if a fourth-dimensional object were cut crosswise its section would be a cube; that is, a four-dimensional object

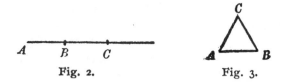

Fig. 2.   Fig. 3.

is bounded on all sides by solids. Again, on a line we can find two points equidistant from each other; for example, the points B and C with the single distance BC (Fig. 2). In a plane, we can find three equidistant points, as the vertices of an equilateral triangle in which $AB = BC = CA$ (Fig. 3). In our space, four equidistant points can be located, the vertices of a

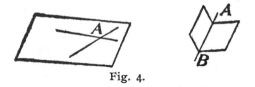

Fig. 4.

tetrahedron, that is, a pyramid having four triangular faces. Hence, in four-dimensional space it should be possible to find five equidistant points. Further, rotation in a plane takes place about a point; in our space, about an axis, as shown in Fig. 4. Hence, in four-dimensional space, rotation should take place about a plane.

This last point—rotation—leads to a curious geometrical application of the principle. We have in Fig.
5 two triangles, of which the sides and angles of the
one are equal to the corresponding sides and angles
of the other. We can lift one triangle up and turn
it over on the other so that the two triangles fit
exactly together. But, mind, we could not do it other-

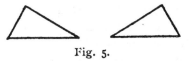

Fig. 5.

wise than by lifting. Hence, these two triangles could
never be fitted together by the mathematicians of flatland, since to them lifting is inconceivable. Possibly,
however, they might suspect this method by noticing
that an inhabitant of one-dimensional space—say, for
simplicity, one living on a straight line—might experience a similar difficulty in comparing the equality of

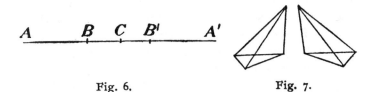

Fig. 6.   Fig. 7.

two segments, AB and $B'A'$ (Fig. 6), each defined
by a set of two points. We may suppose that the segments are equal and so that the corresponding points
in them could be superposed by rotation round C.
This movement, so simple to a flatlander, would be
inconceivable to our one-dimensional being. In fact,
even if he were moving along the lines from A to A',

he would not arrive at the corresponding points in the same relative order, and thus might hesitate to believe that the corresponding distances were equal. So, judging from this being's difficulties, the dweller of the plane might infer, by analogy, that by turning one of the triangles over through three-dimensional space they could make them coincide.

We have a somewhat similar difficulty in our geometry. Let us suppose two pyramids (Fig. 7) similarly related. All the faces and angles of the one correspond exactly to the faces and angles of the other. Yet lift them about as we please, we could never fit them together. If we fit the bases together, the two will lie on opposite sides, one being below the other. Again, we may conceive of two solids, such as a right hand and a left hand, which are exactly similar and equal, but of which one cannot be made to occupy exactly the same position in space as the other does. These are difficulties similar to those experienced by the inhabitants of flatland in comparing the triangles. But it may be conjectured that in the same way as such difficulties in the geometry of an inhabitant in space of one dimension are explicable by moving the figure temporarily into space of two dimensions by means of rotation round a point, and as such difficulties in the geometry of flatland are explicable by moving the figure temporarily into space of three dimensions by means of rotation round a line, so such difficulties in our geometry would disappear if we could temporarily move our figures into space of four dimensions by means of rotation round a plane—a movement quite inconceivable to us. That is, the dweller in four-dimensional space would take our troublesome pyramids and fit them together without any trouble. By merely turning over one of them he would convert it into the other without any change whatever in the relative positions of its parts. What he could do with the pyramids he could also do with our hands or our right shoe and left shoe, or, in fact, with one of us human beings, if we allowed him to take hold of us and turn a somersault with us in the fourth dimension. We should then return to our own space and appear as changed as when our natural form is seen in a mirror. Everything on us would be changed from right to left, even the seams of our clothes, and every hair on the head. And through the whole process no change would occur in the relative positions of the parts of the body.

To sum up, then, we may say that the fourth dimension is an extension in a space unknown to us and in a direction outside of those we can conceive. The idea is to us incomprehensible. We have no positive proof of its existence. But, inconclusive and insufficient as are the results, we can get a notion of the fourth dimension by attending to the corresponding step that the plane being would have to take in forming an imaginary construction of our space. Also, we considered how this inhabitant of flatland might find arguments to support the view that space of three dimensions existed, and then we saw whether analogous arguments applied to our world. Right around us, but in a direction that we can no more conceive than the flatlander can conceive a direction perpendicular to his plane, there may exist, then, another universe, or any number of universes. All that physical science can say against this supposition is, that even if a fourth dimension exists, it must forever remain unknown to us in our natural condition.

In conclusion, it may be said that the growth of this "fairyland of geometry" has been greatly influenced by the theory of parallels, which theory is the result of an attempt to prove that through a point only one line can be drawn parallel to another line—this being taken for granted in our plane geometries. Ignoring, then, this accepted truth, Lobachevsky, a Russian geometer, and the Hungarian Bolyai constructed, about 1830, a geometry in which more than one line can be drawn through a point parallel to another line. The term applied to it is "metageometry," and its study has stimulated the development of the geometry of hyperspace, of which the fourth dimension is but a special case.

Furthermore, attempts have been made to find, in the space of four dimensions, explanations of certain difficulties or apparent inconsistencies in physical science. Thus, the behavior of the atoms in certain carbon compounds has been attributed to their motion in the fourth dimension. Attempts have also been made to explain the properties and constitution of matter by means of space of four dimensions. One writer has even argued thus: If a finite solid were passed slowly through flatland, the inhabitants would be conscious only of that part of it which was in their plane; that is, they would see only a surface, or the section of the solid by their space. They would see the shape of the object gradually change and finally vanish. In the same way, if a body of four dimensions were passed through our space, we should be conscious of it only as a solid, namely, the section of the body by our space; and as it moved along, we should see its form and appearance gradually change and finally vanish, perhaps. So he suggested that the birth, growth, life, and death of animals may be explained thus, as the passage of finite four-dimensional bodies through our three-dimensional space. Again, the idea of a fourth dimension has been made ridiculous by the suggestion that spirits probably dwell in that dimension and can appear to us and disappear at pleasure, thus offering an explanation for the so-called phenomena of spiritualism. But whatever else we may think of these theories, we can certainly admit the possibility of a fourth dimension, even if it be only for the sake of "mental gymnastics."

XXII.

THE PROPERTIES OF FOUR-DIMENSIONAL SPACE.

BY "SYLVESTER" (MAJOR WILMOT E. ELLIS, COAST ARTILLERY CORPS, U. S. A.).

Dimension as applied to space signifies extension.
These extensions are measured in directions mutually perpendicular to one another, and the number of dimensions is determined by the number of independent perpendicular directions that can exist in the given space.

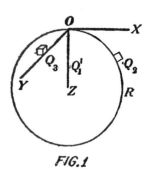

FIG. 1

To illustrate: In Fig. 1, assume that ZO is a line drawn from the center of a sphere to the surface. All points in the line, from an assumed origin, O, to infinity in either direction could be represented by giving different values to *one* variable, as z, and using the proper sign, plus or minus.

Imagine that the line ZO is the only space in existence, and that a mathematical intelligence is concentrated in the point Q_1. It would have one sense of direction only, an "up-down" sense, for it could form no conception of any motion perpendicular to its line. From these two premises (the one algebraical, and the other geometrical), it follows that a line is a one-dimensional space.

At the point O erect OX perpendicular to OZ. A plane passed through these two lines will cut a circle ZOR from the sphere. The mathematical intelligence in this case may be represented by the square Q_2. We may imagine it confined in the plane to the immediate proximity of the circumference of the circle ZOR, just as our habitat is located in the immediate vicinity of the surface of the earth. Q_2, however, has one more space perception than Q_1, for the former, in traveling its circumference, adds a sense of "forward-backward." Q_2 can move in either direction, OZ or OX. These two directions may be assumed at pleasure in the plane, but having assumed arbitrarily any one direction, only one other perpendicular direction can exist in the plane. Every point in the plane of ZOX may be reached by giving proper values to two variables, x and z. Hence, a plane is a space of two dimensions.

If, at the point O, we draw a perpendicular OY to the plane of ZOX, we determine a new space of three dimensions. The mathematical intelligence, now represented by the cube Q_3, has the added perception of "right-left."* This is "our" "solid" space. The essential characteristic of this space is that, at each point, any number of three independent perpendicular directions may be determined, but no more than three. All points in "our" space may be located by giving different algebraical values to three variables, as x, y, and z.

Let us assume that at the point O a fourth line, OW, could be drawn perpendicular to the three axes, OX,

---

*In this discussion, the "up-down" sense, associated with the attraction of gravitation, has been assumed as our primary sense of direction, because most of our physical perceptions are either directly or indirectly referred to gravitative force. It should be remembered, nevertheless, that the order of development of the three senses is immaterial, as the gravitative direction has no significance in geometry.

OY, and *OZ*. We should thus determine a four-dimensional space, and the mathematical intelligence, Q_4, dwelling therein, would have a new perception of direction, which, for the lack of a better name, we may call the "w" sense. We cannot represent the "w" direction in a figure, nor Q_4 by any known geometrical form. Since every line whatsoever in "our" space may be regarded as belonging to some set of axes, it follows that the "w" direction must be perpendicular to all lines in three-dimensioned space, in effect perpendicular to the space itself.

We are absolutely lacking in the "w" sense. The key to this direction is concealed from the mathematical genius as well as from the schoolboy. The question naturally arises: Is this limitation a human limitation only, or is there something inherent in what we might term "absolute reason" that precludes the idea of the fourth and higher dimensions? It is at least possible that the limitation exists in human reason alone. General geometry, both pure and analytical, ascends from zero to any number of dimensions without any break to betray the passing of the third dimension.

If we accept a "w" direction, definitely abandoning all hope of mentally representing it, we can investigate the properties of four-dimensioned space as satisfactorily as we can those of three dimensions. Our only method of investigation must be analogy, but we shall find that it will not once fail us. Following this line of inquiry, we may develop the following properties of four-dimensioned space:

1. A line includes an infinity of points, or zero-spaces; a surface, an infinity of lines, or 1-spaces;*

---

* For brevity, 1-space, 2-space, etc., will be frequently used to signify one-dimensioned space, two-dimensioned space, etc.

and a solid, an infinity of surfaces, or 2-spaces. We are justified in concluding, therefore, that a 4-space includes an infinity of 3-spaces. A 3-space is but one of many in a 4-space, and a fourth-dimensional intelligence would view "our" 3-space as an insignificant part of his 4-space.

2. In analytical geometry, it is shown that any point in 1-space can be represented by an equation of the general form $x = a$; a line in 2-space by the general equation $ax + by = c$; and a plane in 3-space by the general equation $ax + by + cz = d$. So in 4-space, a 3-space may be represented by the general equation $ax + by + cz + dw = e$.

3. In 2-space, three points can be so located as to have any arbitrary distance between pairs of points; in 3-space, four points can be so located; and in 4-space, five points. As illustrations, for conditions of equal distances, we have the equilateral triangle in 2-space, and the regular tetrahedron (regular pyramid), in 3-space.

4. As a line is generated by the motion of a point, a surface by the motion of a line, a solid by the motion of a surface, so a fourth-dimensional body may be generated by some motion of a solid.*

5. A polygon is bounded by three or more lines; a polyhedron, by four or more polygons, and a fourth-dimensional body by five or more polyhedrons.

6. In 2-space, rotation can take place only about a point; in 3-space, about a line; and in 4-space, about a plane.

---

*Mathematicians have demonstrated that in 4-space, there should be six regular structures corresponding to the five regular polyhedrons of 3-space. For example, the analogue of the cube is bound by 8 cubes, with 16 corners, 24 squares, and 32 edges. These structures can only be vaguely conceived by the most imaginative mathematicians.

7. Two geometrical magnitudes are said to be symmetrical, when every point of the one has a corresponding point at the same distance on the opposite side of an assumed spatial reference. The symmetry here defined is what is known as two-fold. It is not necessary to consider other kinds of symmetry. In one dimension, symmetry exists with respect to a point; in two dimensions, with respect to a line; in three dimensions, with respect to a plane. An object and its mirror reflection are always symmetrical. Such figures are equal, but to prove their coincidence, it is necessary to turn one of them around, "upside down," or "inside out," as the case may be. This process is called circumversion.

8. In order to circumvert a figure, it must be turned around or maneuvered in the next higher dimension. Thus, a line must be turned through a plane, a polygon through 3-space, and a solid through 4-space.*

Let us assume, for purposes of illustration, that a two-dimensional world and a four-dimensional world has each a separate existence. We must further postu-

---

*If an intelligence capable of visualizing 4-space exists in any realm of the universe, it is more than probable that n dimensions exist for an nth order of intelligence. The eight properties herein postulated of the fourth dimension may be thus generalized, rectilinear figures only being considered

1. An n-space includes an infinity of $(n-1)$ spaces.

2. In a space of n dimensions, an $(n-1)$ space may be represented by an equation of the first degree containing n variables.

3. In an n-space, $n+1$ points may be located so as to have any arbitrary distance between pairs of points.

4. An nth dimensional figure or space may be generated by some motion of an $(n-1)$th dimensional figure or space.

5. An nth dimensional figure is bounded by $n+1$ or more figures of $n-1$ dimensions.

6. In an n-space, rotation can take place only about a space of $(n-2)$ dimensions.

7. In an n-space, symmetry exists with respect to an $(n-1)$ th space.

8. Circumversion in an n-space can be effected only by a movement through an $(n+1)$ space.

late that the 2-world and our world have small, but real extensions in the third and fourth dimensions, respectively. Without these extensions, an imaginary visitor from one world to the next lower could not perform his mysterious feats. Let us also represent concrete mathematical intelligences of the 2-world, 3-world, and 4-world by Q_2, Q_3, and Q_4, respectively. Each of these imaginary beings is supposed to have an intelligence and dimensions corresponding to his own world.

If Q_3 should visit a 2-world, he would be perceived by Q_2 as two-dimensional. For example, if the visitor were cubical in shape, every part would be invisible to Q_2, except the square base of contact. Q_2 could not understand how a coin, "head-up," could be turned "tail-up." Q_3 could easily perform the feat, either by taking the coin into his own space, turning it, and restoring it, or by turning it around a chord of the circular coin as an axis. The maximum element of the circle that could possibly remain visible to Q_2 during the transformation would be a single diameter. Similarly, if Q_4 should visit our world, he would appear as a three-dimensional being. He could turn a sphere "inside out," either by withdrawing it to his own space, or by revolving it through his space around a circle of the sphere remaining in our space.* The maximum element of the sphere that could be seen during any such operation would be a great circle.

Again, if Q_2 were inside the bounding line of any figure, as the circumference of a circle, he could not reach the outside without breaking through. Q_3, by first moving normal to the plane, could pass out and in at will, without penetrating the boundary. So, in

---

* This process does not turn the sphere inside out. See introduction, page 28.

our world, Q_4, by moving in the "w" direction, could pass in and out of a solid sphere without breaking through the surface.

Q_2 could make a simple loop in a string,* but so long as he kept the string intact and the ends fastened, he could not straighten it. Q_3 could do this, however, by lifting the loop into his space, untwisting it, and restoring the string. The corresponding maneuver with us would be the untying of an ordinary ("thumb") knot, without disturbing the fastened ends or cutting the string. Q_3 would have to evoke Q_4 to solve this problem.

It is interesting to note that Slade (who was eventu-

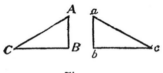

Fig 2.

ally exposed) performed the trick of passing a grain of corn through the solid surface of a glass sphere, and that of untying a knot, as described in the preceding paragraph. The celebrated mathematician, Zöllner, witnessed these two performances, and appears to have believed that Slade was assisted by fourth-dimensional "spirits."

Q_2 regards the symmetrical triangles of Fig. 2 as different shapes, because he cannot possibly make them coincide. He perceives a disposition of the one with respect to the other exactly analogous to the one we perceive with respect to "right-handed" and "left-handed" shapes. Q_3 proves that the triangles are equal,

---

* See foot-note, page 30.

by moving them until AB and ab coincide, and then rotating one about the line AB-ab, until it falls upon the other. During this rotation, the moving figure is turned "upside down."

If Q_3 views ABC from above the plane of the paper, he obtains one aspect of the triangle. If he views ABC from a point the same distance below the plane of the paper, he obtains the other (the abc) aspect. We see then that Q_3's conception of symmetrical shapes, as but two aspects of one shape, results from his freedom of movement in a direction normal to Q_2's space.

We sense a right glove and a left glove as different shapes, yet we have an intuitive feeling they ought to be the same figures. If one turns a right glove inside out, it becomes a left glove, and vice versa. Q_4 can perform this transformation, when the gloves are closed surfaces. If he rotates a right glove a half-turn through his space, it becomes a left glove, the rotating glove being necessarily turned "inside out" during the movement.*

Suppose that Q_4 should view a glove from two points, the one "above," and the other a corresponding position "below" our space. From one point of view, the glove will appear to him as a right glove, and from the other point of view, as a left glove. He recognizes no difference whatsoever between "outside" and "inside," except one of aspect. It is for this reason that the passage from what *we* call the "inside" of a sphere to the "outside" does not necessarily involve the penetration of the surface. This attribute of Q_4's intelligence results from his ability to conceive and move in what we have called the "w" direction. If I were

---

* See foot-note on page 247. It is not the "outside" and "inside" that Q_4 sees above and below our space, but two new sides.—H. P. M.

ever able to see a right glove as a left glove (except in a mirror), I should know that I possessed a fourth-dimensional intelligence, and could move with perfect freedom in the "w" direction.

It should be noted that if two symmetrical figures be rotated a quarter-turn toward each other through the next higher space, they will coincide. From this point of view, symmetrical figures may be regarded as resulting from a splitting of one figure in a given space, and an unfolding into the next lower space.

With the possible exception of symmetry existing in our world, we have no evidence of the real existence of a 4-world in the finite, and no evidence whatever in the direction of the infinite.

We know that the ether, although it eludes all of our senses, envelopes and permeates our phenomenal world. We feel in some vague, intuitive way, that it is the medium connecting us with a higher order of existence and thought. In the ether, if anywhere, we should expect to find some fourth-dimensional characteristics. Gravitation, electricity, magnetism, and light are known to be due to stresses in, or motions of, the infinitesimal particles of the ether. The real nature of these phenomena has never been fully explained by three-dimensional mathematical analysis. Indeed, the unexplained residuum would seem to indicate that so far we have merely been considering the three-dimensional aspects of four-dimensional processes. As one illustration of many, it has been shown both mathematically and experimentally that no more than five corpuscles may have an independent grouping in an atom; a most significant fact, in view of our third "property" of 4-space.

The fourth dimension has an ethical and philosophical as well as a mathematical and physical value. The idea reveals many fruitful fields of speculation. As examples may be cited the stupendous significance of the first "property" of 4-space, and a pondering of the question: Might not birth be an unfolding through the ether into the symmetrical life-cell, and death the reverse process of a folding-up into fourth-dimensional unity?

A CATALOG OF SELECTED
DOVER BOOKS
IN SCIENCE AND MATHEMATICS

Math–Geometry and Topology

ELEMENTARY CONCEPTS OF TOPOLOGY, Paul Alexandroff. Elegant, intuitive approach to topology from set-theoretic topology to Betti groups; how concepts of topology are useful in math and physics. 25 figures. 57pp. 5⅜ x 8½. 60747-X

COMBINATORIAL TOPOLOGY, P. S. Alexandrov. Clearly written, well-organized, three-part text begins by dealing with certain classic problems without using the formal techniques of homology theory and advances to the central concept, the Betti groups. Numerous detailed examples. 654pp. 5⅜ x 8½. 40179-0

EXPERIMENTS IN TOPOLOGY, Stephen Barr. Classic, lively explanation of one of the byways of mathematics. Klein bottles, Moebius strips, projective planes, map coloring, problem of the Koenigsberg bridges, much more, described with clarity and wit. 43 figures. 210pp. 5⅜ x 8½. 25933-1

CONFORMAL MAPPING ON RIEMANN SURFACES, Harvey Cohn. Lucid, insightful book presents ideal coverage of subject. 334 exercises make book perfect for self-study. 55 figures. 352pp. 5⅝ x 8¼. 64025-6

THE GEOMETRY OF RENÉ DESCARTES, René Descartes. The great work founded analytical geometry. Original French text, Descartes's own diagrams, together with definitive Smith-Latham translation. 244pp. 5⅜ x 8½. 60068-8

PRACTICAL CONIC SECTIONS: The Geometric Properties of Ellipses, Parabolas and Hyperbolas, J. W. Downs. This text shows how to create ellipses, parabolas, and hyperbolas. It also presents historical background on their ancient origins and describes the reflective properties and roles of curves in design applications. 1993 ed. 98 figures. xii+100pp. 6½ x 9¼. 42876-1

THE THIRTEEN BOOKS OF EUCLID'S ELEMENTS, translated with introduction and commentary by Thomas L. Heath. Definitive edition. Textual and linguistic notes, mathematical analysis. 2,500 years of critical commentary. Unabridged. 1,414pp. 5⅜ x 8½. Three-vol. set. Vol. I: 60088-2 Vol. II: 60089-0 Vol. III: 60090-4

GEOMETRY OF COMPLEX NUMBERS, Hans Schwerdtfeger. Illuminating, widely praised book on analytic geometry of circles, the Moebius transformation, and two-dimensional non-Euclidean geometries. 200pp. 5⅝ x 8¼. 63830-8

DIFFERENTIAL GEOMETRY, Heinrich W. Guggenheimer. Local differential geometry as an application of advanced calculus and linear algebra. Curvature, transformation groups, surfaces, more. Exercises. 62 figures. 378pp. 5⅜ x 8½. 63433-7

CURVATURE AND HOMOLOGY: Enlarged Edition, Samuel I. Goldberg. Revised edition examines topology of differentiable manifolds; curvature, homology of Riemannian manifolds; compact Lie groups; complex manifolds; curvature, homology of Kaehler manifolds. New Preface. Four new appendixes. 416pp. 5⅜ x 8½. 40207-X

History of Math

THE WORKS OF ARCHIMEDES, Archimedes (T. L. Heath, ed.). Topics include the famous problems of the ratio of the areas of a cylinder and an inscribed sphere; the measurement of a circle; the properties of conoids, spheroids, and spirals; and the quadrature of the parabola. Informative introduction. clxxxvi+326pp; supplement, 52pp. 5⅜ x 8½. 42084-1

A SHORT ACCOUNT OF THE HISTORY OF MATHEMATICS, W. W. Rouse Ball. One of clearest, most authoritative surveys from the Egyptians and Phoenicians through 19th-century figures such as Grassman, Galois, Riemann. Fourth edition. 522pp. 5⅜ x 8½. 20630-0

THE HISTORY OF THE CALCULUS AND ITS CONCEPTUAL DEVELOP-MENT, Carl B. Boyer. Origins in antiquity, medieval contributions, work of Newton, Leibniz, rigorous formulation. Treatment is verbal. 346pp. 5⅜ x 8½. 60509-4

THE HISTORICAL ROOTS OF ELEMENTARY MATHEMATICS, Lucas N. H. Bunt, Phillip S. Jones, and Jack D. Bedient. Fundamental underpinnings of modern arithmetic, algebra, geometry, and number systems derived from ancient civilizations. 320pp. 5⅜ x 8½. 25563-8

A HISTORY OF MATHEMATICAL NOTATIONS, Florian Cajori. This classic study notes the first appearance of a mathematical symbol and its origin, the competition it encountered, its spread among writers in different countries, its rise to popularity, its eventual decline or ultimate survival. Original 1929 two-volume edition presented here in one volume. xxviii+820pp. 5⅜ x 8½. 67766-4

GAMES, GODS & GAMBLING: A History of Probability and Statistical Ideas, F. N. David. Episodes from the lives of Galileo, Fermat, Pascal, and others illustrate this fascinating account of the roots of mathematics. Features thought-provoking references to classics, archaeology, biography, poetry. 1962 edition. 304pp. 5⅜ x 8½. (Available in U.S. only.) 40023-9

OF MEN AND NUMBERS: The Story of the Great Mathematicians, Jane Muir. Fascinating accounts of the lives and accomplishments of history's greatest mathematical minds–Pythagoras, Descartes, Euler, Pascal, Cantor, many more. Anecdotal, illuminating. 30 diagrams. Bibliography. 256pp. 5⅜ x 8½. 28973-7

HISTORY OF MATHEMATICS, David E. Smith. Nontechnical survey from ancient Greece and Orient to late 19th century; evolution of arithmetic, geometry, trigonometry, calculating devices, algebra, the calculus. 362 illustrations. 1,355pp. 5⅜ x 8½. Two-vol. set. Vol. I: 20429-4 Vol. II: 20430-8

A CONCISE HISTORY OF MATHEMATICS, Dirk J. Struik. The best brief history of mathematics. Stresses origins and covers every major figure from ancient Near East to 19th century. 41 illustrations. 195pp. 5⅜ x 8½. 60255-9